Elements of Temporal Topos

Goro C. Kato

Published 2013 by abramis

ISBN 978 1 84549 581 7

© Goro C. Kato 2013

All rights reserved

This book is copyright. Subject to statutory exception and to provisions of relevant collective licensing agreements, no part of this publication may be reproduced, stored in a retrieval system, or transmitted in any form or by any means, without the prior written permission of the author.

Printed and bound in the United Kingdom

This book is sold subject to the conditions that it shall not, by way of trade or otherwise, be lent, re-sold, hired out, or otherwise circulated without the publisher's prior consent in any form of binding or cover other than that which it is published and without a similar condition including this condition being imposed on the subsequent purchaser.

abramis is an imprint of arima publishing.

arima publishing
ASK House, Northgate Avenue
Bury St Edmunds, Suffolk IP32 6BB
t: (+44) 01284 700321

www.abramis.co.uk

Preface

A physics theory based on a certain mathematical field is an appropriate theory when the theory can provide clear explanations and explicit formulations for physically empirical phenomena. When the theory gives contradicting implications (or ∞ as a value for limit cases or even worse without the limit cases), such a theory crossed over the range of its useful applicability domain. We are currently in such a situation for general relativity and quantum mechanics applied to, for example, the big bang, black holes, or Planck scale ultra microcosm.

For the study of macrocosm we have Newton mechanics and the general theory of relativity, and for microcosm we have quantum mechanics. However, a theory does not exist showing, for example, the process how the macrocosmic object emerges from a collection of microcosmic objects. A comprehensive understanding of the processes from microcosm objects to the macrocosm object requires explaining how a macrocosm object emerges in an organized manner from the information of micro objects. For example, a macrocosm object M, e.g., a stone or a planet consists of microcosm objects $\{m_k\}_{k \in K}$, e.g., elementary particles. We would like to have an explicit formulation in terms of established mathematical concepts showing that it is indeed the case that each microcosm object m_k is a restriction of the macrocosm object M consisting of the given microcosm objects $\{m_k\}_{k \in K}$. Even more importantly, we would like to explicitly formulate how microcosm objects give an emergence of an organized macrocosm object. In such a formulation attempt, the Uncertainty Principle of Heisenberg needs to be considered when we consider a collection of microscopic objects $\{m_k\}_{k \in K}$ since $\{m_k\}_{k \in K}$ are not fixed background particles forming such a macro object M. Such a theory needs to be a state-sensitive theory, involving observations of $\{m_k\}_{k \in K}$ describing the connections of the states of micro objects and the emerging macro object. Microcosmically speaking, there are no such things as a stable M because of the uncertainty of microcosm objects $\{m_k\}_{k \in K}$. Our choice for formulating such processes is the theory of sheaves. A presheaf is evaluated at an object of a category called a *site*, (See Definition I. 3.1), which play the role of a state determining category for an entity corresponding to the presheaf.

To begin a treatise from the very beginning is sometimes difficult and even dangerous. However, since categories and sheaves are not the common languages for physicists, we begin with a concise introduction of a category. In order to show our fundamental and direct approaches to micro and macro physical entities, we begin with notions from a functor and a natural transformation. Even though a certain knowledge of pure mathematics might be useful for our approach, we assume only that the reader is able to think abstractly in terms of categorical concepts. Having some knowledge in abstract algebra, functional analysis, complex analysis, and topology might be helpful, but we do not assume such a background for the reader when introducing our formulations of our theory of temporal topos. In this treatise, the only mathematical background needed for temporal topos is systematically exposed.

The author expresses his thankfulness to Christine Willis for her English language comments on the manuscript of this treatise. The author also thanks Prof. dr. van der Merwe for suggesting the title Elements of Tomporal Topos.

Finally, it is an honor to dedicate this treatise to

Professor Alwyn van der Merwe.

                                                                    Goro C. Kato, v. l. s.

Acknowledgement

The author is grateful for the invitations from the following institutes and universities and the following mathematicians and physicists who shared their thoughts: Institute for Advanced Study, the University of Rochester, UC, Irvine, the University of Antwerp, Imperial College, Lund University, Darmstadt University of Technology, Indian Statistic Institute, George Mason University, Calabria University, Research Institute for Mathematical Sciences (RIMS) of Kyoto University, Institute of Economic Research (IER) of Kyoto University, Okayama University, Tsukuba University, Josai University, Shizuoka University, Waseda University, Tohoku University, and P. Deligne, R. Langlands, (late) A. Weil, S. Lubkin (my Ph.D. advisor), D. Prill, (late) L. Nachbin, F. van Oystaeyen, D. Larsson, I. Raptis, K. E. Wolff, S. Roy, M. Kafatos, D. C. Struppa, T. Kimura, T. Kawai, H. Kashiwara, M. Sato, K. Nishimura, H. Nakamura, T. Kogiso, Y. Baba, K. Kondo, H. Asashiba, A. Koyama, K. Hashimoto, Y. Morita, S. Takemae, T. Tanaka.

Table of Contents

    Preface   1
    Prologue   4

Chapter I   Categories   5

    Section I.1   Categories and Functors   5
    Section I.2   Yoneda Lemma   13
    Section I.3   Sites, Coverings, Sheaves over a Site, and
                        Decompositions of Presheaves   17

Chapter II   Cohomology of Sheaves for Physics; Abstract Differential Geometry,
                Twister Covering Cohomology, and p-Adic String   25

    Section II.1   Derived Functors   25
    Section II.2   Cohomologies via Coverings   29
    Section II.3 D-Modules   33

Chapter III   Temporal Topos   46

    Section III.1 Associated Presheaves and Space-Time Presheaves   46
    Section III.2 Particle-Wave Duality and Ur-Entanglement   60
    Section III.3 Micromorphisms, Uncertain Principle and
                        Quantum Tunneling   65
    Section III.4 Microdecompositions and Microcoverings;
                        Fundamental Presheaves and Particles   75
    Section III.5 Limits, t-Entropy, u-Singularities, Light Cones, and
                        Black Holes   78

References   91

Index   95

Prologue: Nature is surrounded by chaos; however, if we can focus on the common essence of events by ignoring the trivial, an underlying principle may be found. For example, the earth is going around the sun, and the moon is going around the earth. Ignoring the superficial, what is a shared principle (if and when it exists) behind the events? The notion of gravity is the hardest force to be microcosmically handled.

Mathematical formulations capturing the fundamental principles of nature has been one of the main activities in modern science. One of the main questions is what is an appropriate model or formulation representing the basic principles of nature, ideally presenting only the essence of nature. As we have been witnessing in the history of science, mathematics has been proved to be useful for describing the fundamental principles of nature. For describing the fundamental roles in quantum and relativistic views, we can list mathematical fields like group theory, Hilbert space (or more generally functional analysis), differential geometry, and algebraic geometry. As a candidate for a unifying theory of the microcosm and the macrocosm, one insists on (A): the mathematical consistency as a necessary condition. Next, (B): the naturalness of the theory may be required even though the judgment for the naturalness of such a theory could be a subjective matter. Lastly, (C): simplicity of the theory would be preferable. Our temporal topos theory is an attempt to meet these three requirements (A), (B), and (C).

Our choice of such a mathematical language is coming from categories and sheaves, which originally developed independently of motivation to be used for physics. The temporal topos theory is simple since the only concepts appearing in this theory are objects, morphisms, and functors (presheaves) between categories. Temporal topos is efficient since notions in our temporal topos cover from microcosm phenomena to macrocosm phenomena even though we will introduce various notions using those simple concepts in category theory. Finally, temporal topos is natural since each entity like a macro particle is associated with a presheaf whose decompositions into subpresheaves correspond to yet smaller particles composing the macro particle. Such fundamental characterizations for microcosm as particle-wave duality, Heisenberg uncertainty relation among physical observables, and quantum entanglement are essentially embedded in the temporal topos theory. Furthermore, temporal topos theory is capable of describing light cones and applications to cosmology in macrocosm and in microcosm.

It is hoped that *temporal topos theory* gives a structural frame which is universal in the sense of explaining the fundamental principles of both microcosm and macrocosm. Our method of the quantum theory of space and time affected by mass in micro and macro scales is hoped, as naturally as possible, to be transformed into the language of temporal topos, in the transition from microcosm to macrocosm. Namely, our temporal topos (t-topos) is capable of treating mass affected space-time with one theory without changing mathematical models for the process from microcosm to macrocosm.

Chapter I Categories

Section I.1 Categories and Functors

The theory of categories was originally developed by S. Eilenberg and S. MacLane in the mid 1940's [15]. Since then, category theory together with sheaf theory has been effectively applied particularly to algebraic geometry, complex analytic geometry (i.e., the theory of holomorphic functions of several complex variables in H. Cartan Seminars in 1950's), and algebraic analysis (especially in the theory of $D-$modules in 1970's Sato School at the Research Institute for Mathematical Sciences, Kyoto University).

In physics, Christopher Isham (and the Theoretical Physics Group at the Imperial College) is the leading physicist in the study of quantum gravity in terms of sheaves, i.e., topos. His fundamental papers [30] and [12] are important and inspirational papers. An even earlier paper by Chris Isham in 1993, *"Prima Facie Questions in Quantum Gravity"* is motivational for developing a new quantum gravity theory in terms of sheaves.

The various categorical concepts including functors and natural transformations provide devices to accurately express intuitive images of nature. The notions of inverse and direct limits are to play an important role in describing the various aspects of *(u-) singularities*. For our treatment of space and time, another crucial notion comes from a variant form of Yoneda's Lemma. A concise description of categorical methods leading to those concepts will be given in this chapter. When a theory is developed enough to speak for itself, the role of capturing nature is reversed. Namely, one needs to listen to the theory for what it says about nature. That is, the devices to describe nature are telling us what the theory is telling us about nature. It is our approach that nature is captured in the "ur-world" in terms of presheaves.

One of the orthodox ways to logically introduce the notion of a category may be to begin with the notion of a *universe* as in [35]. A reader who is interested in such a logical introduction to category theory is recommended to read [35] and [21] with a connection to algebraic analysis and algebraic geometry. In this treatise, we will ignore any set theoretic logical issues. Hence, our approach to introducing categories is a practical one so that the concept of *temporal topos (t-topos)* can be defined immediately. That is, even though more elegant and general presentations of categorical notions are known, we focus on the physically applicable presentations for our t-topos. Namely, we shall keep close contact with the applications of our *t-topos* connections associated with physical meanings. We also provide motivations and reasons for introducing various concepts in category theory. As a physicist looks for mathematics fields as useful tools to describe physical phenomena, one often needs to modify the established mathematical fields to be a physical theory. As we shall see, *t-topos* is not an exception. As a starting point, the notion of a *topos* of *presheaves* over a *site* is a good model for us to use, and later we will need a certain degree of modification for our *t-topos* theory.

Fundamentally speaking, the set of axioms of a category has the fewest requirements on a collection of

"things" (objects) and their "relations" (morphisms)

which are shared by many fields in mathematics. In these mathematically oriented preparatory chapters, we will often embed our physical applications that are studied in the following chapters. Our basic approach is to grasp the physical events through the information flows, i.e., morphisms of a category. First of all, we begin with the definition of a category since categorical notions are not common knowledge for physicists.

Definition I.1.1 A category $\mathfrak{A}$ consists of
(C.1) The set $Ob(\mathfrak{A})$ of elements which are said to be *objects* of category $\mathfrak{A}$.
(C.2) For an ordered pair of objects A and B of $\mathfrak{A}$, we have a set $Hom_{\mathfrak{A}}(A,B)$. An element f of $Hom_{\mathfrak{A}}(A,B)$ is said to be a *morphism* from A to B, written as

$$f: A \longrightarrow B \text{ or } A \xrightarrow{f} B.$$

Note that sets $Hom_{\mathfrak{A}}(A,B)$ and $Hom_{\mathfrak{A}}(A',B')$ are disjoint unless $A = A'$ and $B = B'$.

(C.3) Let A, B and C be objects of $\mathfrak{A}$. For $f: A \longrightarrow B$ and $g: B \longrightarrow C$, there exists the composition morphism $g \circ f: A \longrightarrow C$.

(C.4) The composition in (C.3) is associative, i.e., for any $A \xrightarrow{f} B \xrightarrow{g} C \xrightarrow{h} D$ in $\mathfrak{A}$, we have $h \circ (g \circ f) = (h \circ g) \circ f$.

(C.5) For each object A of $\mathfrak{A}$, there exists a morphism $1_A \in Hom_{\mathfrak{A}}(A,A)$ satisfying $f \circ 1_A = f$ and $1_A \circ g = g$ for any $f: A \longrightarrow B$ and $g: C \longrightarrow A$.

We have completed the definition of a category. Namely, a category consisting of the set $Ob(\mathfrak{A})$ and the set $\bigcup Hom_{\mathfrak{A}}(A,B)$ where $A,B \in Ob(\mathfrak{A})$ satisfies conditions (C.1) through (C.5). Next we will make remarks on the definition of a category.

Remarks I.1.2 (1) For $f: A \longrightarrow B$, object A is said to be the *domain* of f, and object B is said to be the *codomain* of f.

(2) Notice that $1_A \in Hom_{\mathfrak{A}}(A,A)$ is unique since $1_A = 1'_A \circ 1_A = 1'_A$ for another such $1'_A \in Hom_{\mathfrak{A}}(A,A)$.

(3) A morphism $f: A \longrightarrow B$ is said to be an *isomorphism* if there exists a morphism $g: B \longrightarrow A$ so that we have $f \circ g = 1_B$ and $g \circ f = 1_A$.

The first example of a category is to regard a topological space as a category. This is the classical case on which the classical sheaf theory, i.e., before the 1960's, is based. However, we are going to later define a *presheaf* as a *contravariant functor* from a category to the category of sets. Namely, we need the general concept of a *site* for our t-topos theory. That is, we need more morphisms between objects than just an inclusion mapping. This is one of the reasons why we need a general notion of a site rather than just a topological space.

Example I.1.3 (1) Let T be a topological space. We define a category T associated with the topological space T as follows. The set $Ob(T)$ of objects of T consists of open subsets U, V, --- of the topological space T. We define the set of morphisms between U and V of $Ob(T)$ as follows. Set $Hom_T(U,V)$ is an empty set if U is not contained in V, and if U is contained in V, the set $Hom_T(U,V)$ consists of one morphism, i.e., just the inclusion map from U to V. Then one can confirm that T becomes a category.

(2) Let Ab be the category of abelian groups. Namely, an element of $Ob(Ab)$ is an abelian group, and an element of $Hom_{Ab}(A,B)$ is an abelian group homomorphism. Similarly, let (Set) be the category of sets. Then an object of (Set) is a set, and a morphism of (Set) is a set-theoretic mapping. Other examples are the category of rings whose objects are rings and whose morphisms are ring homomorphisms, and the category of modules over a ring R (as a special case, the category of vector spaces over a field, e.g., the field of complex numbers, or real numbers) whose objects are modules and whose morphisms are R-linear abelian group homomorphisms.

(3) For a category C, consider a category C^{opp} whose objects $Ob(C^{opp}) = Ob(C)$ and $Hom_{C^{opp}}(A,B) = Hom_C(B,A)$. Category C^{opp} is said to be the *dual category* of C. That is, a morphism $f: A \longrightarrow B$ in C becomes $f^{opp}: B \longrightarrow A$ in C^{opp}.

(4) Let C and C' be categories. Then category C' is said to be a subcategory of C when $Ob(C')$ is contained in $Ob(C)$ satisfying the following. For any objects A and B of C', we have $Hom_{C'}(A.B) \subseteq Hom_C(A.B)$, and the composition of morphisms in C' is the restriction of the composition of C to C'. We also say that C' is a full subcategory of C when $Hom_{C'}(A,B) = Hom_C(A,B)$. The category of finite dimensional vector spaces over a field k is a full subcategory of the category of infinite dimensional vector spaces over the field k since the notion of a linear transformation (k-linear mapping) is irrelevant to the dimensions.

(5) In t-topos, each (physical) quantity measurement is assigned a category where an observation of the quantity is made. Since we consider a finite number of quantities to be measured in what will follow, we need to consider a finite product of categories. Here is the definition of a product of two categories. One can easily generalize the notion of product to an arbitrary finite number of categories. The *product category* $C \times C'$ of categories C and C' is defined by the following. An

object of the product category $C \times C'$, i.e., an element of $Ob(C \times C')$, is an ordered pair (A,A') of objects $A \in Ob(C)$ and $A' \in Ob(C')$. A morphism from (A,A') to (B,B') of such pairs of $Ob(C \times C')$ is a pair (f,f') of morphisms of C and C' where $f: A \longrightarrow B$ and $f': A' \longrightarrow B'$ in C and C', respectively.

(6) An object K in a category C is said to be a *terminal object* of C, if for each object A in C, $Hom_C(A,K)$ consists of a unique element. An *initial object* of C is a terminal object in the dual category of C. Let K' be another terminal object of C. Then we have a unique morphism $K \xrightarrow{\sigma} K'$. On the other hand we also have a unique morphism $K' \xrightarrow{\sigma'} K$. Then the composition of $K \xrightarrow{\sigma} K' \xrightarrow{\sigma'} K$ is a morphism $K \xrightarrow{id_K} K$. The uniqueness implies $id_K = \sigma' \circ \sigma$ and similarly $id_{K'} = \sigma \circ \sigma'$. Namely, any two terminal objects in C are isomorphic. Hence one can call a terminal object "the terminal object" by identifying isomorphic terminal objects. When an object is both terminal and initial, such an object is said to be a *zero object* of C.

Definition I.1.4 A morphism $f: A \longrightarrow B$ in category C is said to be a monomorphism if $f \circ g = f \circ g'$ implies $g = g'$ for all pairs of morphisms g and g' with codomain A. Another way of saying this is: for any object D of C, the mapping $Hom_C(D,A)$ to $Hom_C(D,B)$ defined by g to $f \circ g$ is a set theoretic injective. One can say that a morphism $A \longrightarrow B$ is determined by the composition with a morphism from an arbitrary object D to A which is relevant to Remark I. 2. 2. (4). Similarly, we define an epimorphism $f: A \longrightarrow B$ if the induced mapping sending g to $g \circ f$, i.e., $Hom_C(B,D) \longrightarrow Hom_C(A,D)$, is injective. Furthermore, $f: A \longrightarrow B$ is said to be an *isomorphism* when there exists a morphism $f': B \longrightarrow A$ satisfying $f \circ f' = 1_B$ and $f' \circ f = 1_A$. If $f: A \longrightarrow B$ is an isomorphism, then f is both a monomorphism and an epimorphism. But in general the converse is false.

Definition I.1.5 Let A be an object of category C. Let $g: S \longrightarrow A$ and $g': S' \longrightarrow A$ be monomorphisms. Then write $S \geq S'$ if there exists a morphism $\alpha: S' \longrightarrow S$ to satisfy $g \circ \alpha = g'$. Then monomorphisms g and g' are said to be *equivalent* if both $S \geq S'$ and $S' \geq S$ hold. That is, for such equivalent monomorphisms, such an $\alpha: S' \longrightarrow S$ becomes an isomorphism from S' to S. The definition of a *subobject* of A is an equivalence class of a monomorphism $g: S \longrightarrow A$. Dually, we define a *quotient object* of A in C as follows. By defining equivalent epimorphisms from A to Q and to Q', we get an isomorphism $\beta: Q \longrightarrow Q'$. Namely, a quotient object of an object A is a subobject in the dual category C'. A *subquotient* object is an extremely significant concept. We will define homological and cohomological objects as subquotient objects in Chapter II.

The concepts of inverse and direct limits are important for our t-topos. For t-topos theory, we need to discuss those limits for a functor from a site (a category with *a Grothendieck topology*) to a category. The intuitive ideas of inverse and direct limits may be interpreted as the far left-side object and the far right-side object, respectively, of a sequence in a category C of the form

$$---\longrightarrow F(V)\longrightarrow F(U)\longrightarrow ---,$$

where V and U are objects in an indexing category C', and F is a functor from C' to C. There are several ways to define inverse and direct limits. We choose more direct and intuitive definitions as follows. For other ways to introduce the notions of limits, one can consult [42], [35], and other treatises.

Definition I.1.6 Let F be a covariant functor from C' to C. Namely, for $\alpha_j^i : i \longrightarrow j$ in C' we have $F\alpha_j^i : Fi \longrightarrow Fj$, or $F\alpha_j^i : F_i \longrightarrow F_j$. An inverse limit in C, written as $\varprojlim_i F_i$, of the system $F\alpha_j^i : F_i \longrightarrow F_j$ is an object of C satisfying the following.

(Inv.1) For each i in C', there exists a morphism $F_i \longleftarrow \varprojlim_i F_i$ in C such that diagram

$$\begin{array}{ccc} & \varprojlim_i F_i & \\ \swarrow & & \searrow \\ F_i & \xrightarrow{F\alpha_j^i} & F_j \end{array}$$

is commutative, and the following universality is satisfied.

(Inv.2) If there exists another such object Y as to satisfy the condition in (Inv.1), there exists a morphism $Y \longrightarrow \varprojlim_i F_i$ satisfying the obvious commutative diagram.

As the dual notion, for a system $F\alpha_j^i : F_i \longrightarrow F_j$, one can also define the *direct limit* $\varinjlim_i F_i$ as an object of C in a similar way. Namely, for $\alpha_j^i : i \longrightarrow j$ in C', we have the commutative diagram

$$\begin{array}{ccc} & \varinjlim_i F_i & \\ \nearrow & & \nwarrow \\ F_i & \xrightarrow{F\alpha_j^i} & F_j. \end{array}$$

For any object Y in C, which satisfies the above commutative diagram, there exists a unique morphism from $\varinjlim_i F_i$ to Y. That is, the direct limit $\varinjlim_i F_i$ is the initial object among all those objects Y satisfying the above commutativity.

Next we will define the image of a morphism. The image of a morphism is interpreted as the information of an observation (or measurement) in the t-topos approach.

Note that the notion of a *germ*, e.g., in complex function theory, is defined by the direct limit of functions as follows. Let f and g be holomorphic (analytic) functions defined over domains U and V, respectively, in a complex plane $\mathbb{C}$ so that $U \cap V$ is not empty. We often write $f \in O(U)$ and $g \in O(V)$. Let z be a point in $U \cap V$. Define an equivalence relation $f \sim g$ as follows. For $z \in W \subset U \cap V$, if f and g coincide on W, namely the restrictions of f and g on W are equal, i.e., then we write $f \sim g$. Then the equivalence class f_z is said to be the germ of f at z. Classically, the germ f_z is a *function element* in the sense of Weierstrass.

Definition I.1.7 Let $f: A \longrightarrow B$ be a morphism in a category C. The *image*, written as $\mathrm{Im}\, f$ of f is a subobject of B. That is, the image of f is a monomorphism $\iota: \mathrm{Im}\, f \longrightarrow B$ satisfying $f = \iota \circ f'$ for some $f': A \longrightarrow \mathrm{Im}\, f$ together with the following universal mapping property. If any object I satisfies the same property as $\mathrm{Im}\, f$, i.e., if $\iota': I \longrightarrow B$ is a monomorphism such that $f = \iota' \circ f''$ for $f'': A \longrightarrow I$, then there exists a monomorphism $\iota'': \mathrm{Im}\, f \longrightarrow I$ satisfying the canonical commutativity, i.e., $f'' = \iota'' \circ f'$ and $\iota = \iota' \circ \iota''$. This definition of the image $\mathrm{Im}\, f$ of $f: A \longrightarrow B$ indicates that $\mathrm{Im}\, f$ is the smallest subobject of B among those subobjects I in the above definition acting like an image of $f: A \longrightarrow B$.

The concept of a *functor* is crucial and most fundamental not only for t-topos, but also for any approach using category-sheaf theory in physics. In what will follow in t-topos theory, one associates a particle with a *presheaf* defined on an appropriate category called a *temporal site (t-site)*. As we will explain in detail in Section III. 1, this t-site plays the role of a state controlling parameter for the particle.

Definition I.1.8 Let C and C' be categories. A *covariant functor* F from C to C' is defined as follows: for an object A of C, FA (or $F(A)$) is an object of C', and for a morphism $f: A \longrightarrow B$ in C, we have a morphism $FA \xrightarrow{Ff} FB$ in C'. Then for the compositions, $F(g \circ f) = Fg \circ Ff$ must hold where $A \xrightarrow{f} B \xrightarrow{g} C$ is in C and $FA \xrightarrow{Ff} FB \xrightarrow{Fg} FC$ in C'. Furthermore, for the identity $1_A \in \mathrm{Hom}_C(A, A)$, we must have the preservation; the induced morphism by functor F on the identity

$F1_A : FA \longrightarrow FA$ is indeed the identity morphism of FA, i.e., $1_{FA} : FA \longrightarrow FA$. That is, we have $F1_A = 1_{FA}$.

A *contravariant functor* F can be defined dually. Namely, for $f : A \longrightarrow B$, we have $Ff : FB \longrightarrow FA$ satisfying $F(g \circ f) = Ff \circ Fg$ and $F1_A = 1_{FA}$. Notice that a contravariant functor from C to C' is a covariant functor from the dual category C^{opp} of C to C'.

This completes the definition of a covariant (contravariant) functor from category C to category C'. We call C and C' the *domain category* and the *codomain category* of F, respectively. As mentioned earlier, our definition of a presheaf is simply a contravariant functor from category C to the category of sets. Before we give examples of functors, we will give the definition of a morphism of functors.

Definition I. 1. 9 Let C and D be categories and let F and G be contravariant functors from C to D. A *morphism* α *of functors* from F to G is defined as follows:

(F.1) For each object A of C, we have a morphism $\alpha_A : FA \longrightarrow GA$.

(F.2) For $f : A \longrightarrow B$ in C, the contravariant functors F and G induce the morphisms $Ff : FB \longrightarrow FA$ and $Gf : GB \longrightarrow GA$. Then the diagram

$$\begin{array}{ccc} FB & \xrightarrow{Ff} & FA \\ \downarrow & & \downarrow \\ GB & \xrightarrow{Gf} & GA \end{array}$$

is commutative, i.e., $\alpha_A \circ Ff = Gf \circ \alpha_B$.

A morphism of functors is classically said to be a *natural transformation*. Such a morphism α of functors is said to be an isomorphism when α_A for each A in C is an isomorphism in C. Then such a morphism (natural transformation) α of functors is said to be a *natural isomorphism* (or *natural equivalence*).

Note I.1.10 (1) From Definition 1.9, we can consider the category whose objects are (contravariant) functors and whose morphisms are morphisms of functors (natural transformations) in the sense of the above Definition 1.9. We write such a category as D^C. Note that for a functor $C' \xrightarrow{F} C$, we have the induced functor $F^* : D^C \longrightarrow D^{C'}$, which for an functor $\varphi \in Ob(D^C)$, assigns $\varphi \circ F \in Ob(D^{C'})$. See the following diagram:

$$\begin{array}{ccc} C & \xrightarrow{F} & C' \\ \downarrow & \swarrow \varphi & \\ D & & \end{array}.$$

Similarly for a functor $D' \xrightarrow{G} D$, we get the induced functor $G_* : D^C \longrightarrow (D')^C$ from the diagram

$$\begin{array}{c} C \\ \downarrow \searrow \\ D \xrightarrow{G} D'. \end{array}$$

When the above D is the category (Set) of sets, we also write $\hat{C} = (Set)^C$. Notice that, e.g., an object of the category of abelian groups Ab has an element; however, an object of $\hat{C} = (Set)^C$ does not have an element.

(2) Let C and D be categories and let $F : C \longrightarrow D$ be a covariant functor between those categories. For each $f : A \longrightarrow B$ in C, we get the induced morphism $Ff : FA \longrightarrow FB$ in D. Namely, we have a mapping of sets

$$Hom_C(A,B) \longrightarrow Hom_D(FA,FB).$$

Then F is said to be an *equivalence* of categories if and only if the mapping $Hom_C(A,B) \longrightarrow Hom_D(FA,FB)$ is a bijection (injective and surjective), and F is essentially surjective in the following sense: for any object in D, there exists an isomorphic object of the form FA in D. Notice that an equivalence $F : C \longrightarrow D$ of categories can be phrased as follows: there exists a functor $F' : D \longrightarrow C$ so that $F' \circ F$ and $F \circ F'$ are isomorphic to 1_C and 1_D, respectively, as morphisms of functors as in Definition 1.9. Proving this equivalent assertion is left for the reader. When the set theoretic mapping $Hom_C(A,B) \longrightarrow Hom_D(FA,FB)$ defined by $f \in Hom_C(A,B) \longrightarrow Ff \in Hom_C(FA,FB)$ is injective (surjective), F is said to be *faithful (full)*. When F is both faithful and full, F is said to be *fully faithful*.

Next we will give examples of functors.

Examples I.1.11 (1) Let C be a category and let (Set) be the category of sets. Then, $Hom_C(-,B)$ is a contravariant functor from C to (Set). This is because for an object A of C, we get an object of (Set), i.e., the set $Hom_C(A,B)$ of morphisms from A to B. Furthermore, the contravariantness of the functor $Hom_C(-,B)$ can be observed by the following diagrams. Namely, for

$$A \xrightarrow{f} A'$$

in C, we get the morphism in the following direction in

$$Hom_C(f,B) : Hom_C(A',B) \longrightarrow Hom_C(A,B).$$

The induced mapping $Hom_C(f,B)$ by $Hom_C(-,B)$ is defined as follows: for h in $Hom_C(A',B)$, we have $h \circ f$ in $Hom_C(A,B)$. We now introduce convenient notation for the functor $Hom_C(-,B)$. Namely, define $\tilde{B} = Hom_C(-,B)$. Then $Hom_C(f,B)$ becomes $\tilde{B}(f)$. On the other hand, for an object A of C, $Hom_C(A,-)$ becomes a covariant functor from C to (Set).

(2) We will consider a global section functor later when we are ready to introduce the notion of a sheaf. This global section functor will be shown to be a covariant functor from the category of sheaves to a category, e.g., to the category Ab of abelian groups. (As was shown, *direct* and *inverse limits* are also extremely important functors for t-topos theory and are some of the main functors appearing in t-topos theory.)

Next we are going to consider a functor which can be represented by an object of the category on which such a functor is defined. Namely, we begin the following definition.

Definition I. 1.12 Let F be a contravariant (or covariant) functor from C to (Set). Then F is said to be a *representable functor* if $\tilde{B} = Hom_C(-,B)$ is isomorphic to F as a morphism of functors where B is an object of C. The object B is said to be a *representing object* for F. Namely, there is an isomorphism (i.e., a bijection of sets) $\alpha_A : Hom_C(A,B) \longrightarrow FA$ for every object A of C. That is, using the terminology in Definition 1.9, when there exists a natural isomorphism (natural equivalence) from $\tilde{B}$ to F, F is said to be a representable functor. Tautologically speaking, $\tilde{B} = Hom_C(-,B)$ is a representable presheaf.

Section I. 2 Yoneda's Lemma

Now we are ready to assert a natural and important lemma for not only our theory but also for the foundations of category-sheaf theory. In our temporal topos, the consequence of Yoneda's Lemma is crucial to treat the presheaves of space, time and space-time. We will give the formulation of space and time, or space-time based on Yoneda's Lemma. We are going to give various views of Yoneda's Lemma because of its importance in our theory. Historically speaking, capturing the true or more accurate nature of space-time has been one of the main issues not only in physics but also in philosophy.

The meaning of Yoneda's Lemma is the following. In the above definition of a representable functor, we have the isomorphism

$$\alpha_A : Hom_C(A,B) \xrightarrow{\approx} FA,$$

or as a morphism of functors in $\hat{C} = (Set)^C$ there is an isomorphism

$$\tilde{B} \xrightarrow{\approx} F.$$

The question is whether the representing object B can be replaced by F itself or not. We cannot have a morphism $\tilde{F} \xrightarrow{\approx} F$ since $\tilde{F}$ and F belong to different categories. However, in order for $\tilde{F} \longrightarrow F$ to make sense, we evaluate at objects of C so that the "values" are in (Set). Let A be an object of C and let $\tilde{A}$ be object of $\hat{C} = (Set)^C$. Then $\tilde{F}\tilde{A} \longrightarrow FA$ makes sense since $\tilde{F} = Hom_{\hat{C}}(-, F)$ needs the object $\tilde{A}$ of $\hat{C} = (Set)^C$ to be evaluated. Therefore, $\tilde{F}\tilde{A} \xrightarrow{\approx} FA$ reads $\tilde{F}\tilde{A} = Hom_{\hat{C}}(\tilde{A}, F) \xrightarrow{\approx} FA$, that is, the assertion of Yoneda's Lemma. In this sense, Yoneda Lemma is an extremely natural assertion.

Yoneda's Lemma I.2.1 Let C be a category and let F be a (contravariant) functor from C to (Set). We have a bijection between the two sets in the above, i.e.,

$$\tilde{F}\tilde{A} = Hom_{\hat{C}}(\tilde{A}, F) \xrightarrow{\approx} FA.$$

For the above bijection between those sets, first let $r \in \tilde{F}\tilde{A} = Hom_{\hat{C}}(\tilde{A}, F)$, i.e., r is a morphism of functors $r: \tilde{A} \longrightarrow F$. In particular, at $A \in Ob(C)$, we have $r_A : \tilde{A}A \xrightarrow{\approx} FA$. Then for the identity $1_A \in \tilde{A}A = Hom_C(A, A)$, we get $r_A(1_A) \in FA$. Define a mapping $\alpha : Hom_{\hat{C}}(\tilde{A}, F) \xrightarrow{\approx} FA$ by $\alpha(r) = r_A(1_A)$. We will show that α is a bijection. On the other hand, we define a mapping $\alpha' : FA \longrightarrow Hom_{\hat{C}}(\tilde{A}, F)$ as follows. For an element of x of FA, we need an element in $Hom_{\hat{C}}(\tilde{A}, F)$. Namely, for this x, we need to assign a morphism s of functors $\tilde{A} \longrightarrow F$. For an arbitrary object of C, define $s_Y : \tilde{A}Y = Hom_C(Y, A) \longrightarrow FY$ as follows. For $f \in Hom_C(Y, A)$, we get $Ff : FA \longrightarrow FY$. This mapping Ff gives an element of FY by defining $Ff(x) \in FY$. Hence, the definition of $s : \tilde{A} \longrightarrow F$ is given by $\alpha'(x)_Y(f) = s_Y(f) = Ff(x)$. The commutative diagram

$$\begin{array}{ccc} \tilde{A}A = Hom_C(A, A) & \longrightarrow & FA \\ \downarrow & & \downarrow \\ \tilde{A}Y = Hom_C(Y, A) & \longrightarrow & FY \end{array}$$

implies that those correspondences are bijective. More explicitly, for 1_A in $\tilde{A}A$, we have $Ff(\alpha(r)) = \alpha'(\alpha(r)_Y(f)$. Next the counter-clockwise diagram chasing for 1_A can be done as follows. First, we have $Hom_C(f, A)(1_A) = f \in \tilde{A}Y$. For $r \in Hom_{\hat{C}}(\tilde{A}, F)$, the above commutative diagram implies that $r_Y(f) = (Ff)(r_A(1_A))$ for arbitrary Y in C and $f \in \tilde{A}Y$. See [42] or [35] for details.

We give several consequences of Yoneda Lemma:

Remarks I.2.2 (0) First of all, Yoneda's Lemma as it is stated, should be read:

(YL) *Elements* in set $F(A)$ can be replaced (isomorphically) by *morphisms of functors* from $\tilde{A}$ to F in $\hat{C}$. Consequently, together with Yoneda's Embedding, one can replace objects and morphisms with presheaves (contravariant functors) and morphisms of functors (i.e., natural transformations).

By the embedding result in the following (2) of this remark, we can replace $\tilde{A}$ with A in (YL). If we use the notion of a scheme in algebraic geometry, the above formulation is read as follows. The set $FA = F(A)$ can be regarded as A-rational point on scheme F where F is a scheme over a commutative ring A. Then $FA = F(A)$ is regarded as the set of common zeros of a given system of equations (an algebraic variety). Then (YL) says that A-rational points on F are one to one correspondence with morphisms from (affine) scheme A to scheme F. We regard this as another face of naturalness of Yoneda's Lemma.

(1) In a tautologically sounding way, Yoneda's Lemma says that F is representing F itself. This is actually correct in the following *lifted sense*. Let us recall first that a representing object is an object in C so that very covariant (or contravariant) functor F can be represented as the lifted object $\tilde{A}$ in $\hat{C} = (Set)^C$ of A in C.

(2) In Yoneda's Lemma, replace F with $\tilde{B}$, where as before B is a representing object of F. Then the equation becomes

$$\tilde{\tilde{B}}\tilde{A} \stackrel{def}{=} Hom_{\hat{C}}(\tilde{A},\tilde{B}) \longrightarrow \tilde{B}A \stackrel{def}{=} Hom_C(A,B).$$

Then this *covariant functor* ~ is a fully faithful functor from C to $\hat{C} = (Set)^C$ as defined in (2) of Note 1.10. Then functor ~ is also said to be an *embedding* in the following sense. The definition of an embedding functor is that the correspondence induced by the functor between the categories is injective both on morphisms and objects. Namely, in addition to the above injective correspondence on morphisms $Hom_C(A,B) \longrightarrow Hom_{\hat{C}}(\tilde{A},\tilde{B})$ (our functor ~ is even an isomorphism), the functor ~ is also injective on objects. In our case, in order to show functor ~ to be injective on objects, we need to show that $\tilde{B} = \tilde{B}'$ implies $B = B'$. For an arbitrary object A in C, $\tilde{B}A \stackrel{def}{=} Hom_C(A,B) = Hom_C(A,B') \stackrel{def}{=} \tilde{B}'A$ implies $B = B'$ by the definition of a category, i.e., by (C.2) of definition 1.1. Namely, the assignment

$$\sim : C \longrightarrow \hat{C},$$

defined by $B \longrightarrow \tilde{B}$, is an embedding which is called *Yoneda's embedding*. We often identify an object with the embedded object in algebraic and topological senses. If one identifies objects and morphisms with the images under an embedding, we may write them as equalities, for example, $\tilde{B} = B$ and $\tilde{f} = Hom_C(-,f) = f$ and also for the embedding from $\hat{C}$ to $\hat{\hat{C}}$ we have $\tilde{\hat{F}} = F$. Namely, we have $---- = \tilde{\tilde{F}} = \tilde{F} = F$ for any object F in an arbitrary category by the Yoneda embedding. Consequently, for any functors F and G in $\hat{C}$, an equation like FG can be read as $F(G)$ as done in algebraic geometry in the following sense:

$$F(G) = \tilde{F}(G) = Hom_{\hat{C}}(-,F)(G) = Hom_{\hat{C}}(G,F).$$

We consider a flexible consequence as the above equation to be an important corollary of Yoneda's Lemma and its embedding for our t-topos theoretic study of space-time. Namely, we are going to replace F with the *sheaf ω associated with space-time,* i.e., as $F = \omega = (\kappa, \tau)$, and G with a particle presheaf m. See Definition I. 3.2 for the definition of a sheaf. The sheaf associated with space-time ω is a *terminal object* in $\hat{C}$. Namely, for any presheaf m in $\hat{C}$, we have a unique morphism of functors $\sigma_m : m \to \omega$. We use κ and τ for the sheaves associated with space and time, respectively, in what will follow. One of the consequences of capturing space-time as a terminal object of $\hat{C}$ is that, as we shall see in Definition III. 1. 1., one cannot observe space-time without the morphisms $\sigma_m : m \to \omega$ from all the associated particle presheaves m in the t-topos $\hat{S}$. For example, an observation $\omega \xrightarrow{s} P$ of the space-time ω, for the presheaf m associated with a particle, by P induces the observation of m by composing the unique morphism $\sigma_m : m \to \omega$ with a measurement morphism $\omega \xrightarrow{s} P$. In a later section, we will study the interconnections of such a morphism σ from m to ω and an observation (measurement) morphism s from ω to P with the notion of *light cones*. The notation in earlier papers of the space-time sheaf (as a terminal object) associated with a particle presheaf m is ω_m which is $\omega(m) = \omega m$. Note that a morphism from a presheaf to a sheaf is by definition a morphism of presheaves, i.e., a morphism in the category $\hat{C}$.

(3) According to the definition of an equivalence of categories as we mentioned in Note 1.10. (2), a category C is equivalent to the subcategory $\tilde{C}$ of representable functors in $\hat{C}$. In Note 1.10, (2), let D be the category $\tilde{C}$ of representable functors, i.e., the images of $\sim$ from C to $\hat{C}$.

(4) The last comment on Yoneda's Lemma is that for presheaves F and G as in (2) of this Remark, to give a morphism α of functors from F to G in $\tilde{C}$ is to give a morphism α_A in C from FA to GA. Yoneda's Lemma gives the following interpretation. For an arbitrary $f \in \tilde{F}\tilde{A} \cong FA$, we get $\alpha \circ f \in \tilde{G}\tilde{A} \cong GA$. As a diagram in $\hat{C}$, under the identification A with $\tilde{A}$, we get the diagram

$$F \xrightarrow{\alpha} G$$
$$\uparrow \nearrow$$
$$A \quad .$$

Namely, for presheaves F and G, any object in any category always can be lifted to a functor so that

(Y.E.P.) *a morphism of presheaves is determined by the composition with a morphism from an arbitrary object in that category.*

For sheaf ω associated with space-time, there is a unique morphism σ_m from a particle associated presheaf m to ω. The measurement s by P of ω is determined by the composition $s \circ \sigma_m$. See Remark I. 3. 3, (3) in Section 1.3 for a connection with the above Yoneda Embedding Principle. As in Chapter III, we will replace the general $\hat{C}$ with a t-topos $\hat{S}$ and define the notion of a measurement between presheaves in $\hat{S}$ as a morphism of contravariant functors which is defined in Definition I. 1. 9. Via the unique morphism σ_m from any presheaf m to the terminal (space-time sheaf) object ω of $\hat{S}$, space-time sheaf ω is playing a role as a measuring sheaf of all the objects in $\hat{S}$. Then Yoneda's Lemma plays a crucial role in the t-topos theoretic formulation of space-time: for our t-topos theory, the importance is that space-time sheaf $\omega = (\kappa, \tau)$ is regarded as a terminal object in $\hat{S}$, and for an arbitrary m in t-topos $\hat{S}$ we have

$$\omega(m) = Hom_{\hat{S}}(m, \omega) = \tilde{\omega}(m).$$

See Chapter III, Section 1 for more details. We shall call the totality of the consequences from Yoneda's Lemma and Embedding *"Little Zen of Yoneda."* See the summary given in Section I. 3.

Note I. 2. 3 For an additive category C and an additive functor F from C to the category of abelian groups, the Yoneda isomorphism is a group isomorphism. There is a generalization of Yoneda's Lemma where the category of sets is replaced as the codomain category of a functor, by a closed category. See [72] for the generalization.

Section I. 3. Site, Coverings, Sheaves over a Site, and Decompositions of Presheaves

For our *t-topos* theoretic approach, objects of the domain category for presheaves play the role of parameters for the states of a particle. In order to distinguish an entity whether the entity is in an ur-particle state or in an ur-wave state, we will not use an expression like "Let p be a particle at time t." We keep track of not only each state of a particle but also each state of time (and space) with an object of the domain category. Such a domain category is said to be a *t-site* which we will define. In a way, our approach says that each particle (hence also the associated time and space in the sense of Yoneda's Lemma formulation) has a uniquely determined state whenever it is observed. The interconnections among space-time formulated in terms of Yoneda's Lemma, light cones, and measurements will be exposed fully in what will follow in Chapter III.

We begin with the definition of a site. A site plays the role of a topological space for the classical sheaf theory. As already indicated earlier, we need more morphisms than one between two objects (classically open sets) in a site for our application. One morphism may correspond to the usual linearly ordered time and others may be corresponding to morphisms even outside light cones or simply covering morphisms. An object of the site (later it will be called the t-site with a more restricted sense than a general site) can be regarded as a parameter for the state of a particle and space-time. For example, the presheaf τ associated with time is also parameterized by an object of the t-site. An object of a t-site is said to be a *generalized time period*. We are avoiding the *Dedekind-Cantor continuum* concept in physics entities.

Definition I.3.1 A *site* S is a category S with a *Grothendieck topology*. A Grothendieck topology on S is defined as follows. For each object U of S, there corresponds a set of morphisms $\{U \longleftarrow U_i\}$ which is said to be a covering of U. Then such a set $\{U \longleftarrow U_i\}$ of morphisms must satisfy the following axioms:

(G1) For an isomorphism $V \xrightarrow{\sim} U$, the set $\{U \xleftarrow{\sim} V\}$ consisting of one morphism is a covering of U.

(G2) When $\{U \longleftarrow U_i\}$ is a covering of U, and $V \xrightarrow{f} U$ is an arbitrary morphism, then the fiber product $U_i \underset{U}{\times} V$ exists, and the pull-back $\{V \longleftarrow U_i \underset{U}{\times} V\}$ is also a covering of V by the projections. See the diagram:

$$\begin{array}{ccc} U_i \underset{U}{\times} V & \longrightarrow & U_i \\ \downarrow & & \downarrow \\ V & \xrightarrow{f} & U. \end{array}$$

(G3) Let $\{U \xleftarrow{g_i} U_i\}$ be a covering of U and let $\{U_i \xleftarrow{g_{ij}} U_{ij}\}$ be a covering of each U_i. Then $\{U \xleftarrow{g_i \circ g_{ij}} U_{ij}\}$ is a covering of U. Namely, a covering of a covering is a covering.

First let us begin with the notation of the exactness of the following type of a sequence in (*Set*)

$$A \xrightarrow{f} B \begin{array}{c} \xrightarrow{g} \\ \xrightarrow[h]{} \end{array} C.$$

The exactness of the above sequence means that f is a monomorphism, and the image of f is exactly where g and h agree. Namely, $\mathrm{Im}\, f = \{b \in B : g(b) = h(b)\} \subset B$.

Definition I. 3. 2 A presheaf F defined over a site S is said to be a *sheaf* if $\{U \xleftarrow{g_i} U_i\}$ is an arbitrary covering of an object U of S. Then the following sequence is exact:

$$FU \xrightarrow{Fg_i} \prod_i FU_i \begin{array}{c} \xrightarrow{\bar{p}_1} \\ \xrightarrow[\bar{p}_2]{} \end{array} \prod_{i,j} F(U_i \times U_j),$$

where $\bar{p}_k = F(p_k)$ is induced by F from the projections $U_i \times U_j \xrightarrow{p_1} U_i$ and $U_i \times U_j \xrightarrow{p_2} U_j$ for $k = 1, 2$.

Remarks I.3.3 (1) Notice that in the classical topological space case, the above exactness means: if local sections coincide on the overlap of open sets, then there exists a unique global section (since Fg_i is a monomorphism) over U whose restriction to each local datum matches with the given local section.

(2) We will apply this exactness, replacing F in the above sequence with sheaves κ, τ, and ω which are associated with space, time, and space-time, respectively. For any presheaf m associated with a particle, by the lifted form of Yoneda's Lemma we can consider, e.g., $\kappa m = \kappa(m)$ as $\tilde{\kappa} m$. One can consider the above diagram as the following diagram of morphisms to F in $\hat{C}$ by Yoneda's Embedding. See Remark 2.2 for this observation.

$$\begin{array}{ccc} U_i & \longrightarrow & U \\ \uparrow & & \uparrow \\ U_i \times U_j & \longrightarrow & U_j \end{array}.$$

In the above diagram, regarding the morphisms in C as morphisms in $\hat{C}$ makes better sense for our theory. That is, by the observation via Yoneda's Lemma, the above exact sequence for the sheaf condition can be written as follows

$$\mathrm{Hom}_{\hat{C}}(U,F) \xrightarrow{\tilde{F}g_i} \prod_i \mathrm{Hom}_{\hat{C}}(U_i,F) \begin{array}{c} \xrightarrow{\tilde{F}p_1} \\ \xrightarrow[\tilde{F}p_2]{} \end{array} \prod_{i,j} \mathrm{Hom}_{\hat{C}}(U_i \times U_j, F),$$

where we are identifying, e.g., U with the lifted Yoneda embedded $\tilde{U}$ in $\hat{C}$. On the other hand, for an arbitrary F in $\hat{C}$, define the *canonical topology* on C as the set of families of coverings $\{U \leftarrow U_i\}$ insisting that the above sequence

$$Hom_{\hat{C}}(U,F) \xrightarrow{\tilde{F}g_i} \prod_i Hom_{\hat{C}}(U_i,F) \xrightarrow[\tilde{F}p_2]{\tilde{F}p_1} \prod_{i,j} Hom_{\hat{C}}(U_i \times U_j, F)$$

is exact. That is, $\tilde{F} = Hom_{\hat{C}}(-,F)$ is not only a presheaf but also a sheaf. Furthermore, for the covering $\{V \leftarrow U_i \times_U V\}$ of V induced by $V \xrightarrow{f} U$, the corresponding sequence

$$Hom_{\hat{C}}(V,F) \longrightarrow \prod_i Hom_{\hat{C}}(U_i \times V, F) \rightrightarrows \prod_{i,j} Hom_{\hat{C}}(U_i \times U_j \times V, F)$$

is also exact. Namely, any presheaf is a sheaf for the canonical topology. In particular, as noted earlier, a representable functor is a sheaf for the canonical topology. We call such families $\{U \leftarrow U_i\}$ of coverings *universally effective epimorphisms*, i.e., the canonical topology consists of the set of universally effective epimorphisms. See [1].

(3) For an arbitrary presheaf m and, e.g., the space-time associated sheaf ω as the terminal object in $\hat{C}$, we have the following diagram in $\hat{C}$:

$$\begin{array}{ccccccc} & & U_i & & & & \\ & \swarrow & \uparrow & \searrow & & & \\ U & & U_i \times U_j & \rightarrow & m & \xrightarrow{\sigma_m} & \omega \\ & \nwarrow & \downarrow & \nearrow & & & \\ & & U_j & & & & \end{array}$$

where σ_m is the unique morphism associated with the sheaf associated with the terminal object ω. See Remarks III. 5. 8 (1) for a generalization of the above diagram. One could put "~" over, e.g., U in the above diagram if one prefers. But with Yoneda's Lemma and the identification of an object with the embedded object in $\hat{C}$, the diagram becomes meaningful. Since space-time sheaf ω is a sheaf, i.e., by Definition 3.2, those morphisms composed with σ_m from U_i to ω coincide over the "overlaps," i.e., $U_i \times U_j$ can be extended to morphisms from U to ω. The above diagram guarantees the smoothness of space-time obtained from fluctuating micro space-time to macro space-time by the "pasting" property of a sheaf from locally given data (i.e., morphisms from U_i to ω) to the global morphism, i.e., from U to ω. For this very reason, we postulate that space, time, and space-time presheaves are sheaves. In a following chapter, we will focus on sheaf ω associated with space-time.

(4) A presheaf F is said to *behave as a sheaf for* U when the exactness for the sheaf property in Definition I. 3.2 holds for a particular U.

Little Zen of Yoneda: Summary on Sheaf, Presheaf, Site, and Yoneda's Principle

Let C be a site, and as before, let $\hat{C} = (Set)^C$ be the category of presheaves, i.e., the category of contravariant functors from C to the category of sets. Then for $F \in Ob(\hat{C})$, we have an isomorphism of sets $Hom_{\hat{C}}(\tilde{X},F) \xrightarrow{\approx} F(X)$, i.e., Yoneda's Lemma where we define $\tilde{X} = Hom_C(-,X)$. In other words, we get $\tilde{F}(\tilde{X}) \approx F(X)$. For a universally effective epimorphic covering $\{U \longleftarrow U_i\}$, every representable functor is a sheaf. See Remarks I. 3. 3. (2). By Yoneda's embedding, we identify $F \approx \tilde{F} \approx \tilde{\tilde{F}} ---$. The principle of our approach is that we pay more attention to the presheaf represented by an object in the sense of Yoneda embedding than the object. Namely, for morphisms of objects $\{X\}$, we look at the morphisms of presheaves, i.e., contravariant functors $\{\tilde{X}\}$ represented by the objects $\{X\}$. Physical applications of this approach are the following. As we shall see in $(t-1)$ of III. 1, for every particle we associate a presheaf F defined over a site S. (Even though a particle $\bar{m}$ is said to be represented by a presheaf m, we do not mean the categorical representation sense, i.e., not $m = \tilde{\bar{m}}$.) Then, as we have indicated earlier, the usual notation $F(V)$ for an object of V can be viewed as the totality of natural transformations $Hom_{\hat{S}}(\tilde{V},F)$ between the presheaves $\tilde{V}$ and F. Namely, we can live only with (contravariant) functors without objects of a site. Let Π be the product category $\prod_{\beta \in \Lambda} C_\beta$ in Definition III. 1.1. Then we can embed the product category $\Pi = \prod_{\beta \in \Lambda} C_\beta$ into the category of presheaves $\hat{\Pi}$ from Π to the category of sets by Yoneda embedding $\sim$. We have the following diagram of functors and categories

$$\begin{array}{ccc} \uparrow & & \uparrow \\ \hat{\tilde{S}} & \xrightarrow{\tilde{\bar{m}}} & \hat{\tilde{\Pi}} \\ \uparrow & & \uparrow \\ \hat{S} & \xrightarrow{\tilde{m}} & \hat{\Pi} \\ \uparrow & & \uparrow \\ S & \xrightarrow{m} & \Pi, \end{array}$$

where all the vertical functors are Yoneda embeddings, and $\tilde{m} \stackrel{def}{=} Hom_{\hat{S}}(-,m)$ and $\tilde{\tilde{m}} \stackrel{def}{=} Hom_{\hat{\tilde{S}}}(-,\tilde{m})$.

1st Consequence: Every presheaf F is representable in the sense of $F \approx \tilde{F}$, i.e., presented by F itself.

2nd Consequence: For a universally effective epimorphic covering $\{U \longleftarrow U_i\}$, every presheaf, which is representable by the 1st Consequence, is a sheaf, i.e., Remark I. 3. 3, (2).

3rd Consequence: For an arbitrary object X of C, the represented presheaf $\tilde{X}$ is a sheaf for the canonical topology whose coverings are universally effective morphisms. By the embedding, an object of a site with the canonical topology is an object of a topos.

4th Consequence: Since every sheaf F is a presheaf, F is represented by an object X of C, i.e., $F \approx \tilde{X}$. By the identification $\tilde{X} \approx X$, every sheaf can be identified with an object of the site with the canonical topology. With the canonical topology on the site, we can regard $C \approx \tilde{C} \approx \hat{C}$. The identification $C \approx \tilde{C} \approx \hat{C}$ may be interpreted: "Space" where functions are defined, and "Sheaf of (germs of) functions" are exchangeable under the canonical topology on the category C.

Tautology: When a Grothendieck topology is the canonical topology, every presheaf is a sheaf.

This completes the Little Zen of Yoneda, which is referred to as L.Z.Y. in what will follow. For example, the left exact global section functor Γ is defined on the site where the codomain category is the category of sets. When a site is the site induced by a topological space X, the global section functor $\Gamma(X,-)$ from the category of sheaves $\tilde{C}$, i.e., the topos of C, can be interpreted as

$$F(X) = Hom_{\hat{C}}(\tilde{X}, F) = \Gamma(X, F)$$

where $\tilde{X}$ is the terminal object of the topos $\tilde{C}$. The interpretation of sections of sheaf F as $Hom_{\hat{C}}(\tilde{X},F)$ in the above equation is closer to an older fashioned approach to sheaf theory. We will return to the topics on the derived functors of the global section left functor $\Gamma(X,-)$ in Chapter II. Consider the following short exact sequence diagram in $\hat{C}$ from $V \xrightarrow{\lambda} X$.

$$0 \longrightarrow F' \xrightarrow{\iota} F \xrightarrow{\pi} F'' \longrightarrow 0$$
$$\nwarrow \quad \uparrow \quad \nearrow$$
$$V \xrightarrow{\lambda} X$$

where we are identifying the objects of C with the representing presheaves in $\hat{C}$. For example, for $\varphi \in Hom_{\hat{C}}(X,F)$, the composition $\varphi \circ \lambda$ corresponds to the usual restriction map in classical sheaf theory.

One of the important applications of the L.Z.Y. is our treatment of the space-time sheaf, developed in Chapter III. We define the space-time sheaf ω as the terminal object in the category $\hat{C}$ of presheaves. In other words, for any presheaf m, there exists a unique morphism $m \xrightarrow{\sigma_m} \omega$. Space-time sheaf ω can be considered as a functor $\tilde{\omega}$ on $\hat{C}$, i.e., an object of $\hat{\hat{C}}$ as well as an object of $\hat{C}$ itself. Namely, σ_m is the unique morphism in $\tilde{\omega}(m) = Hom_{\hat{C}}(m,\omega)$. For the t-topos theory, $Hom_{\hat{C}}(m,\omega)$ is regarded as the effect of the particle representing presheaf m on space-time ω. Namely, ω acts as a measurement presheaf for all the particles representing presheaves of $\hat{C}$. Then a dynamical aspect is induced when those presheaves are evaluated over generalized time periods determining various states of presheaves involved in such a formulation. We will return to these topics in Chapter III.

As we have mentioned informally, for a particle, we associate a presheaf whose states are parameterized by objects of a (t-) site. When such a particle is decomposable into smaller particles, we need to define a decomposition of a presheaf. This notion of a presheaf decomposition provides a process from macro objects to micro objects. Then, together with the notion of a covering in what will follow, we capture the behaviors and the interplays of macro objects and micro objects in (ur-) wave states.

Definition I.3.4 Let m be a presheaf defined on a category C. Let $\{m_j\}_{j \in J}$ be a finite set of objects of $\hat{C}$. A *product* of $\{m_j\}_{j \in J}$ in $\hat{C}$ is a universal object m with a set of morphisms p_j from m to each m_j satisfying the following axiom:

If for each $j \in J$ there is a morphism q_j from n to m_j, then there exists a unique morphism ι from n to m satisfying $q_j = p_j \circ \iota$. We write $m = \prod_{j \in J} m_j$, and $m = \prod_{j \in J} m_j \xrightarrow{p_j} m_j$ is called the *j*-th projection morphism.

Then, for our approach in Chapter III, $m = \prod_{j \in J} m_j$ is said to be a *decomposition* of m. Furthermore, when each $m_j(V_i)$, if reified, is a microcosmic entity, such a decomposition is said to be a *microdecomposition* (see Section III. 4). Then, each m_j may be said to be a *micro-presheaf*. However, the terminology of a micro-presheaf will not be used in what will follow.

Notes I.3.5 (1) One can notice that the definition of a product in terms of a universal mapping property above is identical to the definition of an inverse limit. Namely, $m = \prod_{j \in J} m_j$ is the inverse limit of $\{m_j\}_{j \in J}$ in Definition I. 1. 6 where there are no morphisms among $\{m_j\}_{j \in J}$ in the case of a product. That is, C', appearing in Definition I. 1. 6, is a discrete category meaning no morphisms in C'. Another consequence of Definition I. 3.4 is that such a product of $\{m_j\}_{j \in J}$ is uniquely determined up to an isomorphism. Hence we can say that m is *the* product of $\{m_j\}_{j \in J}$.

(2) We need not call a covering of an object a decomposition of the covering. Similarly as the notion of a microdecomposition, a covering $\{V \longleftarrow V_i\}_{i \in I}$ is said to be a *microcovering* of V when $m_j(V_i)$ is a microcosm object. Then V_i is said to be a micro generalized time period or simply a micro object of the t-site. See Section III. 4.

(3) In Chapter III, we will define an observation (measurement) morphism as a morphism from the observed to an observer. Let us make a short comment on the relationship between an observation of a part and the whole. Let $m = \prod_{j \in J} m_j$ be a decomposition of, e.g., a macro object m into, e.g., micro objects $\{m_j\}$. When m_j is measured by P over V, we have $m_j(V) \xrightarrow{\alpha_j} P(V)$. (However, the composition of the j-the projection with α_j cannot be defined since $(\prod_{j \in J} m_j)(V)$ need not defined.) This means that, while m_j is measured by P over V, one cannot get a measurement of the global object m itself. Note also that some of the other presheaves $\{m_k\}_{k \neq j}$ of $m = \prod_{j \in J} m_j$ can be in an ur-wave state. One can further decompose m_j as $\prod_{k \in K} m_{jk}$, which will be considered in Section III. 4. Consequently, we obtain a sequence of decomposition as

$$\prod_{j \in J} m_j \longrightarrow \prod_{k \in K} m_{jk} \longrightarrow ---.$$

The direct limit

$$\varinjlim_{j,k,---} (\prod_{j \in J} m_j \xrightarrow{j} \prod_{k \in K} m_{jk} \xrightarrow{k} ---)$$

of the above direct system is relevant to the concept of a fundamental presheaf in Section III. 4 corresponding to elementary particles.

Chapter II Cohomology of Sheaves for Physics; Abstract Differential Geometry, Twister Covering Cohomology, and p-Adic String

Section II. 1 Derived Functors

In this chapter, we give a concise introduction to cohomological methods used in several attempts toward quantum gravity. There are unifying concepts among those attempts which are the notions of sheaf cohomology and the theory of modules over the ring of partial differential operators. Here is the brief history of sheaf cohomology. The first systematic study of cohomologies of sheaves appeared as a seminar series lead by H. Cartan during the 1950's. One of the main motivations of introducing the cohomological approach to the study of holomorphic (analytic) functions of several complex variables is to give an interpretation of K. Oka's fundamental results on Cousin problem. The most famous results are so called Theorems A and B, stating the vanishing of higher cohomology groups with coefficient in the sheaf of germs of holomorphic functions over any domain satisfying a certain convexity condition. Oka himself introduced the similar concept of a sheaf in his work, which is recognized as the sheaf of ideals. Then J. P. Serre adapted and developed Cartan's methods to study algebraic geometry in a modern fashion. However, it was Grothendieck who pushed such cohomological methods to the limit. Such an ultimate approach makes it possible to build a purely categorical cohomology theory. Namely, a topological space is replaced with the already defined concept of a site in Section I.3, and the category of abelian groups is replaced by an abelian category. In what follows, we will develop such a categorically oriented cohomology theory, which we believe to be best for the study of quantum gravity.

Let Ab be the category of abelian groups and let X, Y, and Z be groups, i.e., objects of category Ab. Let us consider morphisms (i.e., group homomorphisms) as follows.

$$X \xrightarrow{f} Y \xrightarrow{g} Z.$$

If we assume the composition is a zero morphism, i.e., $g \circ f = 0$, we have the inclusion: $Im(f) \subset Ker(g)$. Then, the *cohomology group at Y* is defined as the *sub-quotient* group of Y

$$Ker(g) / Im(f).$$

We denote the cohomology at Y by $H^0(X \xrightarrow{f} Y \xrightarrow{g} Z)$.

In what will follow, we will focus on taking cohomologies of such sequences in an abelian category as the value category of a covariant left exact functor of global sections defined on the category of sheaves (or presheaves) as the domain category.

We will give a slightly more general approach to the cohomology theory of sheaves than the methods originally used in the twister theory, abstract differential geometry, and quantum cohomology as in, e.g., [59], [63], [53], [54]. This is because all the newer approaches in quantum gravity have been more general and more abstract than those of the 1970's and 1980's. In addition to such a modern tendency, such a discrete notion of physical entity is ultimately necessary for quantum gravity which we interpret as a refusal of the Euclid-Dedekind-Cantor continuum non-discrete type approach.

First of all, we should be aware that all the cohomologies and homologies currently used in mathematics are understood as special cases of the notion of derived functors. We recommend the following treatises for a deeper study of cohomological algebra: [13] as a classical homological masterpiece, [35] and [5] which are suited for further study in the theory of D-modules, i.e., algebraic (micro) analysis. For Algebraic Geometry, [21] is recommended. The first chapter of [50] is also excellent. See also [42] for an introduction to cohomological methods to algebraic geometry and analysis. For a topological view, [80] is highly recommended.

Before an introduction of the concept of derived functors, we begin the notion of an exact connected sequence of functors. Note that an invariant called urcohomology developed in [27] and [42] is neither an exact connected sequence of functors nor half-exact. However, the self-duality still holds. It is the most general invariant defined for any sequences, which need not be (cochain-) complexes. Because of this generality, applications to physics are expected. Note that in [42] urcohomologies are called precohomologies.

Let C and C' be abelian categories. Technically, one can assume only that C' is an additive category for Axiom (E.C.1). Since we are interested in cohomologies of sheaves (later via a covering which is suited for twister cohomologies), i.e., a left exact functor is evaluated at sheaf, we will consider the covariant case. A system of covariant additive functors from C and C' is said to be an exact connected sequence of covariant functors if the following axioms are satisfied.

(E.C.1) Let $0 \to A' \xrightarrow{\alpha} A \xrightarrow{\beta} A'' \to 0$ be a short exact sequence in C, and for each covariant factor h^i from C to C', there is a "connecting" morphism for $i \geq 0$, $\delta^i : h^i(A'') \to h^{i+1}(A')$. For the following commutative diagram

$$\begin{array}{ccccccccc} 0 & \to & A' & \xrightarrow{\alpha} & A & \xrightarrow{\beta} & A'' & \to & 0 \\ & & \downarrow & & \downarrow & & \downarrow & & \\ 0 & \to & B' & \xrightarrow{\alpha'} & B & \xrightarrow{\beta'} & B'' & \to & 0 \end{array}$$

in C, these connected morphisms must induce the commutative diagram together with functorially induced morphisms:

$$\begin{array}{ccc} h^i(A'') & \xrightarrow{\delta^i_A} & h^{i+1}(A') \\ \downarrow & & \downarrow \\ h^i(B'') & \xrightarrow{\delta^i_B} & h^{i+1}(B') \end{array}$$

for $i \geq 0$. The last axiom is the following.

(E.C.2) For a short exact sequence $0 \to A' \xrightarrow{\alpha} A \xrightarrow{\beta} A'' \to 0$ in C, the induced long sequence
$$0 \to h^0(A') \to h^0(A) \to h^0(A'') \xrightarrow{\delta^0} h^1(A') \to h^1(A) \to --- \text{ in } C' \text{ is exact.}$$

An abelian category C is a category where kernels and cokernels of morphisms in C exist. Assume also that the abelian category C is enough injective. That is, for any object A of C, there exists an injective object I of C so that $A \to I$ is a monomorphism. Then by the exact embedding theorem (See [48] or [56] for such an embedding.) into a subcategory of the category Ab of abelian groups, one is allowed to pick an element in the abelian group, i.e., in the embedded object in Ab. The reader may consider all the notions in the category of abelian groups rather than in an abelian category.

Let $F: C \to C'$ be a covariant additive left exact functor of abelian categories. An object I is said to be an injective object of C when the contravariant functor $Hom_C(-, I)$ is not only a left exact functor but also an exact functor from the abelian category C to the category Ab of abelian groups. This definition of an injective object can be phrased also as follows. For any short exact sequence

$$0 \longrightarrow A' \xrightarrow{\iota} A \xrightarrow{\pi} A'' \longrightarrow 0$$

in category C, the contravariant functor $Hom_C(-, I)$ guarantees the exactness up to the following:

$$0 \longrightarrow Hom_C(A'', I) \xrightarrow{\tilde{\pi}} Hom_C(A, I) \xrightarrow{\tilde{\iota}} Hom_C(A', I).$$

However, when for an injective object I, the induced morphism $\tilde{\iota} \stackrel{def}{=} Hom_C(\iota, I)$ becomes an epimorphism, i.e., we can extend the exactness to

$$0 \longrightarrow Hom_C(A'', I) \xrightarrow{\tilde{\pi}} Hom_C(A, I) \xrightarrow{\tilde{\iota}} Hom_C(A', I) \longrightarrow 0.$$

More explicitly, in the following diagram

$$\begin{array}{ccc}
 & I & \\
{\scriptstyle \lambda}\uparrow & \nwarrow{\scriptstyle \lambda'} & \\
0 \to A' & \xrightarrow{\iota} & A,
\end{array}$$

an arbitrary morphism $\lambda: A' \to I$ can be extended to $\lambda': A \to I$ to satisfy the commutativity $\lambda = \lambda' \circ \iota$, which is exactly the "ontoness" of the induced morphism

$\tilde{\iota} \stackrel{def}{=} Hom_C(\iota, I)$. Notice that an injective object in an abstract category C corresponds to an injective module over a commutative ring R when the abelian category C happens to be the category Mod^R of modules over the commutative ring R. First we are going to define the derived functors of F.

Note II. 1.1. As an application to sheaf cohomology, F is going to be associated with space or an open subset of the space. More concretely for sheaf cohomology, a covariant left exact functor F is the global section functor $\Gamma(X,-)$ over space X.

We define the *n-th right derived functor $R^n FA$ of F evaluated at an object A* of an abelian category C as follows. First take any injective resolution $I^\bullet$ of A. Namely,

$$\begin{array}{ccccccc} 0 & \longrightarrow & I^0 & \xrightarrow{\delta^0} & I^1 & \xrightarrow{\delta^1} & I^2 & \xrightarrow{\delta^2} & --- \\ & & \uparrow & & \uparrow & & \uparrow & & \\ 0 & \longrightarrow & A & \longrightarrow & 0 & \longrightarrow & 0 & \longrightarrow & --- \end{array}.$$

From this injective resolution $A \longrightarrow I^\bullet$ of A, we get the complex $FI^\bullet$ in the abelian category C'. (Note that the sequence $FI^\bullet$ is indeed a complex since we have $F(\delta^n \circ \delta^{n-1}) = F(\delta^n) \circ F(\delta^{n-1}) = 0$ from Definition I. 1. 8.) The definition of the *n*-th right derived functor $R^n FA$ of F at A is the n-th cohomology of the complex $FI^\bullet$. Namely, we have

$$R^n FA = H^n(FI^\bullet)$$
$$= H^n(---\xrightarrow{F\delta^{n-1}} FI^n \xrightarrow{F\delta^n} FI^{n+1} \longrightarrow ---)$$
$$= Ker(F\delta^n) / Im(F\delta^{n-1})$$

As an application to sheaf cohomology, we take the left exact functor F as the global section functor $\Gamma(X,-)$ and take A as a sheaf in the abelian category of sheaves. In this case, we obtain the complex $\Gamma(X,I^\bullet)$ of global sections of the complex $I^\bullet$ of injective sheaves. Namely, this complex $\Gamma(X,I^\bullet)$ is induced by the above injective resolution of A:

$$\Gamma(X,I^\bullet): 0 \to \Gamma(X,I^0) \xrightarrow{\Gamma(X,\delta^0)} \Gamma(X,I^1) \xrightarrow{\Gamma(X,\delta^1)} \Gamma(X,I^2) \xrightarrow{\Gamma(X,\delta^2)} ---.$$

The *n*-th derived functor $R^n\Gamma(X,-)A$ of the left exact covariant functor $\Gamma(X,-)$ evaluated at the sheaf A is defined as the *n*-th cohomology $H^n(\Gamma(X,I^\bullet))$ of the above complex $\Gamma(X,I^\bullet)$:

$$H^n(\Gamma(X,I^\bullet))$$
$$= H^n(---\xrightarrow{\Gamma(X,\delta^{n-1})}\Gamma(X,I^n)\xrightarrow{\Gamma(X,\delta^n)}---)$$
$$= Ker\Gamma(X,\delta^n)/Im\Gamma(X,\delta^{n-1}).$$

In this case of the left exact global section functor $\Gamma(X,-)$, the n-th derived functor is said to be the *n-th cohomology with coefficient sheaf* A, denoted as $H^n(X,A)$.

Notes II.1.2 (1) Whenever one talks about cohomological methods, the concept of spectral sequences and the concept of a derived category must be mentioned. However, for the purpose of applying cohomologies to physics, we have given the minimum amount of preparation from cohomological algebra.
 (2) In the above, we have defined the notion of right derived functors of a left exact functor. There is a dual notion to this concept which is called the left derived functor of a right exact functor, e.g., tensor product. The right derived functors and the left derived functors correspond to the cohomologies and homologies, respectively.

Section II. 2 Cohomologies via Coverings

Since our main focus is sheaf cohomology, a left exact functor is going to be the global section functor from the category of sheaves or from the category of presheaves. We will give a concise description of the notion of a covering cohomology. A standard treatment for Cech cohomology can be found in [21] and [42]. Kashiwara and Schapira's [35] is excellent, however, possibly too general for physics applications. First we will give a definition of a Cech cohomology associated with a covering in a more abstract sense, and next we will give a more practical definition associated with a topological space and the category of abelian groups.
 Let be S a site, i.e., a category with a Grothendieck topology (See Definition I. 3. 1.), and let P_S be the abelian category of presheaves on S with an abelian category A as the codomain category, e.g., the category of abelian groups. When a diagram chasing via elements is needed, we can use the embedding theorem of an abelian category into a subcategory of the category of abelian groups. See [48] or [56] for such embedding theorems and proofs. Since P_S has enough injectives (meaning that for any object there exists a monomorphism into an injective object), we can define the notion of derived functors as developed in the previous Section II.1. Let $\{U \leftarrow U_i\}_{i \in I}$ be a covering of an object U in the site S and let F be a presheaf on S. Then we have the diagram in Definition I. 3.2:

$$F(U) \longrightarrow \prod_i F(U_i) \rightrightarrows \prod F(U_i \times U_j)$$

which would be an exact sequence if F were a sheaf. Let us consider the following functor:

$$\ker(\prod_i F(U_i)) \rightrightarrows \prod F(U_i \times U_j)).$$

We denote the above left exact additive functor as $H^o(\{U \leftarrow U_i\}, F)$. (Notice that for a sheaf F, this kernel $H^o(\{U \leftarrow U_i\}, F)$ would be $F(U)$.) Then we can define the derived functors of this left exact functor $H^o(\{U \leftarrow U_i\}, F)$ as we saw in the previous section. The n-th derived functor $R^n H^o(\{U \leftarrow U_i\}, -)(F)$ of $H^o(\{U \leftarrow U_i\}, -)$ evaluated at presheaf of F is called the $n-th$ Cech cohomology object in the abelian category A for the covering $\{U \leftarrow U_i\}$ evaluated at the presheaf F which is denoted as $H^n(\{U \leftarrow U_i\}, F)$.

On the other hand, we need an equivalent definition which can be used for actual computations of the cohomology groups. Here is another more practical formulation of the Cech cohomologies. For this case, let S be the category induced from a topological space. See Example I. 1. 3. Let F be a presheaf, i.e., an object of P_S having the values in the category A of abelian groups. For a covering $\{U \xleftarrow{g_i} U_i\}_{i \in I}$ of U, let $U_{ij} = U_i \cap U_j$, $U_{ijk} = U_i \cap U_j \cap U_k$, and etc. The presheaf F (a contravariant functor on S) induces the restriction morphism as $\rho^i_{ij} : F(U_i) \longrightarrow F(U_{ij})$ and $\rho^{ij}_{ijk} : F(U_{ij}) \longrightarrow F(U_{ijk})$, etc. Then we get the following sequence:

$$\prod F(U_i) \underset{\rho^i_{ij}}{\overset{\rho^j_{ij}}{\rightrightarrows}} \prod F(U_{ij}) \underset{\rho^{ij}_{ijk}}{\overset{\rho jk_{ijk}}{\underset{\rho^{ik}_{ijk}}{\rightrightarrows}}} \prod F(U_{ijk}) \rightrightarrows \cdots.$$

Then let $d^0 = \rho^j_{ij} - \rho^i_{ij}$, $d^1 = \rho^{jk}_{ijk} - \rho^{ik}_{ilk} + \rho^{ij}_{ijk}$, etc. Then we get the complex $(C^\bullet, d^\bullet)$, where

$$C^\bullet = \{C^j(U_i; F)\}_{j \in \mathbb{Z}^+} : C^0(U_i; F) \xrightarrow{d^0} C^1(U_i; F) \xrightarrow{d^1} C^2(U_i; F) \xrightarrow{d^2} \cdots$$

and where $C^j(U_i; F) = \prod F(U_{i_0 i_1 \cdots i_j})$ and $i_0, ---, i_j \in \mathbb{Z}^+ = \{0,1,2,---\}$. For example, the first two differentials are:

$$d^0((f_i)) = \rho^j_{ij}(f_j) - \rho^i_{ij}(f_i)$$

for $(f_i) \in \prod_i F(U_i)$, and for $(f_{ij}) \in \prod_{i,j} F(U_{ij})$ we have

$$d^1((f_{ij})) = \rho^{jk}_{ijk}(f_{jk}) - \rho^{ik}_{ijk}(f_{ik}) + \rho^{ij}_{ijk}(f_{ij}).$$

Namely, the general d^n of the above complex $C^\bullet$ can be explicitly defined as

$$d^n = \rho^{i_1 i_2 \cdots i_{n+1}}_{i_0 i_1 \cdots i_{n+1}} - \rho^{i_0 i_2 \cdots i_{n+1}}_{i_0 i_1 \cdots i_{n+1}} + \cdots - (-1)^j \rho^{i_0 \cdots i_{j-1} i_{j+1} \cdots i_{n+1}}_{i_0 i_1 \cdots i_{n+1}} + \cdots - (-1)^{n+1} \rho^{i_0 i_1 \cdots i_n}_{i_0 i_1 \cdots i_{n+1}}.$$

Then one can confirm $d^{n+1} \circ d^n = 0$ for all $n \geq 0$. The above complex $C^\bullet$ is said to be the Cech complex. The cohomology of $C^\bullet$, i.e.,

$$H^j(C^\bullet) = Ker(d^j) / im(d^{j-1})$$

is said to be the *j-th Cech cohomology group of the Cech complex $C^\bullet$ induced by the covering $U = \bigcup_{i \in I} U_i$* of the open subset U of a topological space. Then the above cohomology $H^j(C^\bullet)$ of the complex $C^\bullet = \{C^j(U_i;F)\}$ is the j-th derived functor of the 0-th cohomology of the complex $C^\bullet = \{C^j(U_i;F)\}$. Let us compute the 0-th cohomology $H^0(C^\bullet)$ as follows. By definition, the 0-th cohomology $H^0(C^\bullet) = Ker(d^0)$. Note that, for $(f_i) \in \prod F(U_i)$ to be in the $Ker(d^0) = H^0(C^\bullet)$, we have

$$d^0((f_i)) = \rho^j_{ij}(f_j) - \rho^i_{ij}(f_i) = 0 \text{ for } i,j \in I.$$

In particular, if F is a sheaf, there exists a unique f in $F(U)$ satisfying $\rho^U_i(f) = f_i$, where ρ^U_i is the restriction from $F(U)$ to $F(U_i)$. Namely, for a sheaf F, we have

$$H^0(C^\bullet) = Ker(d^0) = \Gamma(U,F) = R^0\Gamma(U,-)(F).$$

The last equality is guaranteed by the fact that the global section functor $\Gamma(U,-)$ is left exact. Furthermore, one can define the higher *Cech cohomology group of the open subset U* by taking direct limit for coverings of U. For a presheaf, we have an isomorphism between the Cech cohomology associated with the covering and the cohomology of the complex $C^\bullet = \{C^j(U_i;F)\}$. Let Sh_S be the category of sheaves over the topological space S. Then we have the following diagrams of categories:

$$Sh_S \xrightarrow{\iota} P_S$$
$$\downarrow \swarrow$$
$$A$$

where ι is the fully faithful functor, i.e., for a sheaf F, $\iota(F)$ is the presheaf forgetting the sheaf property of F. Then for the sheaf F, we have
$H^0(C^\bullet) = H^0(\{C^j(U_i;\iota F)\}) = \Gamma(U,F) = R^0\Gamma(U,-)(F)$. The *Cech cohomology group*, denoted as $\check{H}^j(U,F)$ *with coefficient in the sheaf F*, is defined as the direct limit of the Cech cohomologies associated with coverings where the direct limit is taken over refinements of the coverings of U. That is, a covering $\{U \longleftarrow U'_{i'}\}_{i' \in I'}$ of U is said to be a *refinement* of $\{U \longleftarrow U_i\}_{i \in I}$, when there exists a mapping φ of the index sets from I' to I satisfying $U'_{i'} \subset U_{\varphi(i')}$ for all $i' \in I'$. Furthermore, we have an isomorphism between the Cech cohomology group $\check{H}^j(U,F)$ and the derived functor version of the cohomology with coefficient in the sheaf F in Section II.1, i.e., $H^j(U,F) = R^j\Gamma(U,-)(F)$. See [42] or [21] for related topics and further discussions on cohomologies and coverings.

Then one may consider $\prod_{i,j \in I} F(U_{ij})$ as the space of twister (holomorphic) functions. In order to compute the 1st twister cohomology group, first notice that for $(f_{ij}) \in \prod_{i,j \in I} F(U_{ij})$ to be in the kernel of d^1, $d^1((f_{ij})) = \rho_{ijk}^{jk}(f_{jk}) - \rho_{ijk}^{ik}(f_{ik}) + \rho_{ijk}^{ij}(f_{ij}) = 0$ must hold. This cocycle condition says that the value of the branched contour integral is unchanged by changing the common endpoint of the contours in the intersected region. Notice that the first cohomology group of the twister space is exactly the first cohomology group for this covering $\{U \longleftarrow U_i\}_{i \in I}$. For the 0th cohomology group is simply the kernel of d^0 consisting of elements (f_i) in $\prod_{i \in I} F(U_i)$ for the covering $\{U \longleftarrow U_i\}_{i \in I}$ of U satisfying $d^0((f_i)) = \rho_{ij}^j(f_j) - \rho_{ij}^i(f_i) = 0$. Also note that when each covering element U_i is convex enough (technically speaking, when U_i is holomorphically convex, i.e., being a Stein open set), Cech cohomology of U, after a finite number of refinements (i.e., Cech cohomology of a covering) is isomorphic to the sheaf cohomology defined as the derived functor of the global section.

Remarks II. 2.1 When $\{U \longleftarrow U'_{i'}\}_{i' \in I'}$ is a refinement of $\{U \longleftarrow U_i\}_{i \in I}$, there is the induced morphism by φ of complexes

$$\tilde{\varphi}: C^\bullet = \{C^j(U_i;F)\}_{j \in \mathbb{Z}^+} \longrightarrow C'^\bullet = \{C^j(U'_{i'};F)\}_{j \in \mathbb{Z}^+},$$

where the induced morphism $\tilde{\varphi}$ by φ is defined by
$\tilde{\varphi}(f)_{i_0---i_j} = \rho_{i_0---i_j}^{\varphi(i_0)---\varphi(i_j)}(f_{\varphi(i_0)---\varphi(i_j)})$. Furthermore, $\tilde{\varphi}$ functorially induces the morphism on the covering cohomologies

$$H^j(\tilde{\varphi}): H^j(C^{\bullet}) \longrightarrow H^j(C'^{\bullet}).$$

The above *Čech cohomology group*, denoted as $\check{H}^j(U,F)$, *with coefficient in the sheaf F is defined by the direct limit of those induced morphisms on the covering cohomologies.*

Section II. 3 D-Modules

We give a brief review on the theory of D-modules, i.e., modules over the non-commutative ring of partial differential operators defined on a complex manifold of dimension n. After the review, we will focus on the algebraic methods of the theory of modules over the sheaf of non-commutative rings of partial differential operators with coefficients in holomorphic functions. Let X be an n-dimensional complex manifold and let $(z_1, z_2, ---, z_n)$ be local coordinates at a point $z \in X$. Let D_X be the sheaf of non-commutative rings of holomorphic partial differential operators. Namely, at $z \in X$, an element of $P_z \in (D_X)_z$, i.e., a germ P_z in the stalk $(D_X)_z$ at $z \in X$ of the non-commutative sheaf D_X, can be written as

$$P_z = \sum_{\alpha}^{finite} h_\alpha(z)(\partial/\partial z)^\alpha,$$

where the coefficients $h_\alpha(z)$ are (germs of) holomorphic functions, and $\alpha = (\alpha_1, \alpha_2, ---, \alpha_n)$ and $(\partial/\partial z)^\alpha = (\partial/\partial z_1)^{\alpha_1} --- (\partial/\partial z_n)^{\alpha_n}$. Hence, as $P: O_X \longrightarrow O_X$, $P \in D_X$ locally looks like

$$(Pu)(z) = \sum_{\alpha \in \mathbb{Z}^n} h_\alpha(z) \partial_z^\alpha u(z),$$

where $\partial_z^\alpha u(z) = \partial^{\alpha_1 + --- + \alpha_n}(u)/\partial z_1 ---- \cdot \partial z_n$.
Note first that, for example, we have $[\partial/\partial z, f]z = (\partial/\partial z \cdot f - f \cdot \partial/\partial z)z = (\partial f/\partial z) \cdot z$. The ring D_X is a non-commutative ring, and also D_X contains the sheaf O_X of holomorphic functions as a subring. It is known that the sheaf D_X is not only noetherian but also coherent. See [38], [5], [6], or [32]. Any system of l equations of linear partial differential operators P_{ij} for m generators (unknown functions) $u_j, j = 1, 2, ---, m$, can be expressed as

$$\begin{cases} P_{11}u_1 + \cdots - P_{1m}u_m = 0 \\ P_{21}u_1 + \cdots - P_{2m}u_m = 0 \\ - \\ - \\ P_{l1}u_1 + \cdots - P_{lm}u_m = 0 \end{cases}.$$

Such a system can be considered as a germ of a sheaf M over D_X defined over the complex manifold X satisfying the following exact sequence of D_X-modules

$$\cdots \xrightarrow{\cdot Q} D_X^l \xrightarrow{\cdot P} D_X^m \xrightarrow{\cdot u} M \longrightarrow 0,$$

i.e., $\cdot u$ induces the isomorphism of D_X-modules from the 0th cohomology

$$D_X^m / \text{Im}(\cdot P) = H^0(\cdots \longrightarrow D_X^l \xrightarrow{\cdot P} D_X^m \xrightarrow{0} 0)$$

to the D_X-module X.

Remark II. 3.1 Note that the above exact sequence

$$\cdots \xrightarrow{\cdot Q} D_X^l \xrightarrow{\cdot P} D_X^m \xrightarrow{\cdot u} M \longrightarrow 0$$

is a free resolution of the D_X-module M, where $\{u_j\}_{1 \le j \le m}$ is a set of generators for the module M over D_X. The epimorphism $D_X^m \xrightarrow{\cdot u = [u_j]} M \to 0$ is defined by

$P_1U_1 \oplus \cdots \oplus P_m U_m \xrightarrow{\cdot u} P_1 u_1 + \cdots + P_m u_m \in M$, where $U_j = [0, \cdots, \overset{j}{1}, \cdots, 0]$ is the canonical base for D_X^m. Notice also that since D_X^m is noetherian, $\ker(\cdot u)$ is finitely generated over D_X. Hence there exists an epimorphism $D_X^l \xrightarrow{\cdot v} \ker(\cdot u) \longrightarrow 0$. Then the above $\cdot P$ can be expressed as the composition of $\iota: \ker(\cdot u) \longrightarrow D_X^m$ with $\cdot v$. Then we have a free resolution for any finitely generated D_X-module M (i.e., m unknown functions) as above. We are not to give a full cohomological algebra background for the treatment of the theory of D_X-modules here. An interested reader is recommended to read the monographs mentioned above for the interplay between the D_X-module theory and cohomological algebra. If the reader is familiar with the notion of a quasi-isomorphism, the top sequence and the bottom sequence is quasi-isomorphism:

$$---\to D_X^l \xrightarrow{\bullet P} D_X^m \to 0 \to ---$$
$$\searrow \quad \nearrow \iota$$
$$\ker(\bullet u)$$
$$\searrow$$
$$---\to 0 \longrightarrow M \to 0 \to ---\ .$$

That is, cohomologies of the top and the bottom sequences are isomorphic.

Example II.3.2 The sheaf O_X of germs of holomorphic (analytic) functions is an example of a D_X-module. A presentation using an exact sequence is

$$---\longrightarrow D_X^n \xrightarrow{\bullet \partial_{[n,1]}} D_X \xrightarrow{\bullet u} O_X \longrightarrow 0,$$

where $\partial_{[n,1]} = [\partial_1, ---, \partial_n]'$ and $\partial_i = \partial/\partial z_i$ for $i = 1, 2, ---, n$. An explicit presentation as a system of n differential equations with one unknown is

$$\partial_i u = 0,$$

where $i = 1, 2, ---, n$. Note that the quasi-isomorphism can be observed by

$$M \approx D_X / (\partial/\partial z_1, \partial/\partial z_2, ---, \partial/\partial z_n),$$

where $(\partial/\partial z_1, \partial/\partial z_2, ---, \partial/\partial z_n)$ is the ideal generated by the set $\{\partial/\partial z_1, \partial/\partial z_2, ---, \partial/\partial z_n\}$.

For D_X-modules M and O_X as before, let us consider the totality $Hom_{D_X}(M, O_X)$ of D_X-homomorphisms (i.e., -linear homomorphisms) from M to O_X. Then an element of $Hom_{D_X}(M, O_X)$ can be interpreted as a solution of the system of linear partial differential equations represented as the D_X-module M. This is because: for $f \in Hom_{D_X}(M, O_X)$, by using the D_X-linearity of f we have

$$f\left(\sum_{1 \le i \le l} P_{ij} u_j\right) = \sum P_{ij} f(u_j) = \sum P_{ij} f_j = 0,$$

where f_j is the image of u_j under $f \in Hom_{D_X}(M, O_X)$, i.e., holomorphic solutions in O_X of the system satisfying the differential equations. Note that since $f \in Hom_{D_X}(M, O_X)$ is a sheaf homomorphism, it is best to say that f is a local (holomorphic function) solution in O_X of the D_X-module M. If one lets $S = Hom_{D_X}(M, O_X)$, then S is the sheaf of germs of solutions of D_X-module M in

O_X. Then we can consider $Hom_{D_X}(-,O_X)$ as a functor from the category of D_X-modules to the category of sheaves of $\mathbb{C}_X$-modules where $\mathbb{C}_X$ is the constant sheaf of complex numbers. Notice that $Hom_{D_X}(-,O_X)$ is a left exact contravariant functor. Hence we can consider the higher derived functor of the left exact functor $Hom_{D_X}(-,O_X)$. By using the notation of Section II. 1, the n-th derived functor $R^n Hom_{D_X}(-,O_X)M$ of the left exact (solution) contravariant functor $Hom_{D_X}(-,O_X)$ evaluated at the D_X-module M is denoted as $Ext^n_{D_X}(M,O_X)$. Let us compute the 0th and the 1st cohomologies of the general case of a D_X-module M of m unknowns and l equations presented as the exact sequence of D_X-modules:

$$---\xrightarrow{\cdot Q} D_X^l \xrightarrow{\cdot P} D_X^m \xrightarrow{\cdot u} M \longrightarrow 0.$$

Operating the left exact contravariant functor $Hom_{D_X}(-,O_X)$ to this exact sequence, we get

$$Hom_{D_X}(D_X^l,O_X) \xleftarrow{P\cdot} Hom_{D_X}(D_X^m,O_X) \xleftarrow{\tilde{u}} Hom_{D_X}(M,O_X) \longleftarrow 0.$$

Since we have the canonical isomorphism, e.g., $Hom_{D_X}(D_X^m,O_X) \approx O_X^m$, the above sequence becomes

$$0 \longrightarrow S = Hom_{D_X}(M,O_X) \xrightarrow{\tilde{u}\cdot} O_X^m \xrightarrow{P\cdot} O_X^l \xrightarrow{Q\cdot} ---.$$

By the definition of the derived functors as in Section I. 1, we have

$$Ext^0_{D_X}(M,O_X) = R^0 Hom_{D_X}(-,O_X)M = Ker(P\cdot) = H^0(0 \xrightarrow{\tilde{u}} O_X^m \xrightarrow{P\cdot} O_X^l \longrightarrow ---)$$

and

$$Ext^1_{D_X}(M,O_X) = R^1 Hom_{D_X}(-,O_X)M = Ker(Q\cdot)/Im(P\cdot) = H^1(0 \xrightarrow{\tilde{u}} O_X^m \xrightarrow{P\cdot} O_X^l \xrightarrow{Q\cdot} ---)$$

where $P \circ Q = 0$ is referred to as the compatible condition. The global higher solution derived functor is the abutment of the spectral sequence

$$E_2^{p,q} = H^p(X, Ext^q_{D_X}(M,O_X))$$
$$\stackrel{def}{=} R^p\Gamma(X,-)R^q Hom_{D_X}(-,O_X)M$$

and where the abutment is $Ext^{p+q}_{D_X}(X,M,O_{D_X}) \stackrel{def}{=} R^{p+q}(\Gamma(X,-) \circ Hom_{D_X}(-,O_X))M$. Namely, the above spectral sequence is induced by the composition of functors of the global sections and the solution functor. See the following commutative diagram of the categories of D_X-modules, $\mathbb{C}$-modules and abelian groups:

$$(D_X - \mathrm{mod}) \xrightarrow{Hom_{D_X}(-,O_X)} (\mathbb{C} - \mathrm{mod})$$
$$\searrow \qquad \downarrow$$
$$Ab$$

where the vertical functor is the global section functor $\Gamma(X,-)$ and the slanted functor is the composition of those functors. We are not going to develop the theory of spectral sequences here. The interested reader may find the necessary background in the monographs mentioned above.

Remark II.3.3 The meaning of the vanishing of the first cohomology sheaf of (local) solutions in the sheaf O_X of holomorphic functions is the following. Since we have

$$0 = Ext^1_{D_X}(M,O_X) = Ker(Q\bullet)/\operatorname{Im}(P\bullet),$$

we get $Ker(Q\bullet) \subset \operatorname{Im}(P\bullet)$. Namely, for the above sequence

$$0 \longrightarrow Hom_{D_X}(M,O_X) \xrightarrow{\tilde{u}} O_X^m \xrightarrow{P\bullet} O_X^l \xrightarrow{Q\bullet} ---,$$

$f \in Ker(Q\bullet)$ implies $f \in Im(P\bullet)$. That is, f in O_X^l the above sequence satisfying $Qf = 0$ must be written as $Pu = f$ for a certain $u \in O_X^m$.

In addition to the above contravariant solution functor $Hom_{D_X}(-,O_X)$, there is another important functor called the de Rham functor $Hom_{D_X}(O_X,-)$, which is a covariant functor from the category of D_X-modules to the category of sheaves of $\mathbb{C}_X$-modules, i.e., vector spaces over the field $\mathbb{C}$ of complex numbers. With cohomological algebra methods, one can prove the following isomorphism of sheaves

$$Ext^j_{D_X}(O_X,M) \approx H^j(\Omega_X^\bullet \otimes_{O_X} M),$$

whose proof can be found, e.g., in [5], [6], [38], [32]. Notice that the right-hand side of the above isomorphism is the j-th cohomology group of the de Rham type complex $\Omega_X^\bullet \otimes_{O_X} M$. Let us compute a few examples of the de Rham functor $Hom_{D_X}(O_X,-)$. First as a D_X-module M, we consider the following system of equations for one unknown:

$$\partial_k u = 0, \ k = 1,2,---n.$$

Namely, we have the exact sequence, i.e., a free resolution of O_X:

$$D_X^n \xrightarrow{\begin{bmatrix} \partial_1 \\ - \\ \partial_n \end{bmatrix}} D_X \xrightarrow{u} O_X \longrightarrow 0.$$

We compute the higher extension sheaves as

$$Ext_{D_X}^j(O_X, O_X) \approx H^j(\Omega_X^\bullet \otimes_{O_X} O_X) \approx H^j(\Omega_X^\bullet).$$

The far right-hand side $H^j(\Omega_X^\bullet)$ can be computed by holomorphic Poincare Lemma over the complex manifold X as

$$H^j(\Omega_X^\bullet) \approx \begin{cases} \mathbb{C}_X, & j = 0 \\ 0, & j \neq 0 \end{cases}.$$

Another extreme case of D_X-module is D_X itself. That is, D_X-module D_X corresponds to the system of no equations. For this case, i.e., $M = D_X$, we need to compute $Ext_{D_X}^j(O_X, D_X) \approx H^j(\Omega_X^\bullet \otimes_{O_X} D_X)$. We only describe the results:

$$Ext_{D_X}^j(O_X, D_X) \approx H^j(\Omega_X^\bullet \otimes_{O_X} D_X) \approx \begin{cases} 0, & j \neq n \\ \Omega_X^n, & j = n \end{cases},$$

whose proof can be found in [6], [32], [38].

Remarks II. 3.4 (1) For the above computation of the higher extension sheaves, we use the free resolution $\{\Omega_X^j \otimes_{O_X} D_X\}_{j=0,1,---,n}$ of the n-forms Ω_X^n. See [32], [5], [6], and [38] for other functors for the theory of D_X-modules in addition to the above de Rham and the solution functor.
(2) The D_X-module version of *Frobenius Existence Theorem* becomes: The de Rham functor $Hom_{D_X}(O_X,-)$ and the solution functor $Hom_{D_X}(-,O_X)$ induce an equivalence between the category of D_X-modules that are locally free of finite rank as O_X-modules and the category of $\mathbb{C}_X$-modules that are locally free of finite rank (called local systems).
(3) The de Rham functor and the solution functor give an equivalence of categories between the (derived) category of *holonomic and regular* bounded complexes of sheaves of D_X-modules and the (derived) category of bounded complexes of sheaves of $\mathbb{C}_X$-modules that have $\mathbb{C}_X$-constructive cohomologies.

See e.g., [33] for the connection to this ideally generalized version of the 21st problem of Hilbert.

Next we will focus on a special type of D_X-modules. A D_X-module M is said to be an *integrable connection* when M is a free module over O_X. Namely, with a set of generators $\{u_j\}_{1 \leq j \leq m}$, we can write the direct sum $M = \oplus_j^m O_X u_j$.

As an application of the notion of a D_X-module, let X be a topological space and let O_X be a sheaf of abelian groups, i.e., $O_X(U)$ is an abelian group for any open subset U of X. Furthermore, assume that the set $O_X(U)$ of sections of O_X over U is an algebra over a field, e.g., the field $\mathbb{C}$ of complex numbers. When X is a C^∞ differential manifold, $O_X(U)$ is the abelian group of C^∞ functions over U. Let M be an integrable connection. Namely, M is a D_X-module which can be written locally as $M \approx O_X^l$. One can construct the de Rham type complex associated with the locally free O_X-module M:

$$0 \longrightarrow M \xrightarrow{\nabla^0} \Omega_X^1 \otimes_{O_X} M \xrightarrow{\nabla^1} \Omega_X^2 \otimes_{O_X} M \xrightarrow{\nabla^2} ---.$$

We can interpret the above complex $\Omega_X^\bullet \otimes_{O_X} M$ as the $\mathbb{C}_X$-module of the de Rham functor $Hom_{D_X}(O_X, -)$ evaluated at M. For example, see [52], [53], [54], [63] and [64] for the purely algebraic construction of the above complex in terms of abstract differential geometry and its connections to quantum gravity. Note that the complex manifold plays no significant role for the construction of abstract differential geometry. Note also that the (vacuum) Einstein equations for gravity can be obtained from the Riemannian curvature associated with the above de Rham type complex $\Omega_X^\bullet \otimes_{O_X} M$.

Remark II. 3.5 We will give a lengthy remark on the p-adic approaches to string theory by I. Volovich. (See [77] and [78], and the relevant results in the references.) The history of p-adic methods in analysis and (rigid) geometry is long. Especially, p-adic cohomology has been developed for Weil's conjectures on the zeta function associated with an algebraic variety defined over a finite field. Note that such abstract notions currently used in algebraic geometry had been developed mainly by Alexander Grothendieck as a device to prove the conjectures of Andre Weil. See, e.g., [55] and [73] for l-adic cohomology theory and Weil's conjectures. It may not be overly stated that the development in Algebraic Geometry from the 1950's and the 1960's was motivated by Weil's conjectures. As an example of a p-adic approach to physics (p-adic string), developed by the Russian school, we will show how to compute:

(1) the 1st p-adic cohomology group (hence the zeta function) associated with the Fermat curve $X^l + Y^l = 1$ written in the affine form, and

(2) the zeta-endomorphism which is by the Frobenius map on the 1st p-adic cohomology group.

It was first pointed out in [77] that the computation of (2) provides the p-adic Veneziano amplitude. In general, p-adic cohomology is defined as a hypercohomology (hyperderived functor); that is, the coefficients are complexes of sheaves of differential forms whose power series are satisfying a certain growth condition. Hence, it is a type of de Rham cohomology in II. 3. See, e.g., [13], [42], [50], [19], etc. for the notion of a hypercohomology. Let U be the affine Fermat curve as in (1) above defined over the finite field $F_p = \mathbb{Z}/p\mathbb{Z}$, where p is a prime. Let $\underline{U}$ be a lifting of U in the sense of $U = \underline{U} \times_{\hat{\mathbb{Z}}_p} F_p$. The desired cohomology group in (1) is the $\hat{\mathbb{Z}}_p \otimes_{\mathbb{Z}} \mathbb{Q}$-adic 1st cohomology $H^1(U, \hat{\mathbb{Z}}_p \otimes_{\mathbb{Z}} \mathbb{Q})$. We will compute $H^1(U, \hat{\mathbb{Z}}_p \otimes_{\mathbb{Z}} \mathbb{Q})$ as the hypercohomology $H^1(\underline{U}, \Omega^*_{\hat{\mathbb{Z}}_p}(\underline{U})\dagger \otimes_{\mathbb{Z}} \mathbb{Q})$ of the lifting $\underline{U}$ where the †-growth condition mentioned above (called the †-completion) is defined by $\hat{\mathbb{Z}}_p[X,Y]\dagger = \{\sum a_{ij} X^i Y^j ; ord(a_{ij}) \geq \varepsilon(i+j)$ for some $\varepsilon \geq 0$ and for all but finitely many $(i+j) \in \mathbb{N}^2\}$. The †-completion (i.e., the †-growth condition) $\hat{\mathbb{Z}}_p[X,Y]\dagger$ of the polynomial ring $\hat{\mathbb{Z}}_p[X,Y]$ with coefficients in the ring of p-adic integer $\hat{\mathbb{Z}}_p$ is a stronger convergence condition than just a p-adic completion. For more details on the †-completion, see [50], [47] and [42]. The set of generators for this hypercohomology group $H^1(\underline{U}, \Omega^*_{\hat{\mathbb{Z}}_p}(\underline{U})\dagger \otimes_{\mathbb{Z}} \mathbb{Q})$ is

$$\{X^\alpha Y^{\beta-l+1} dX\}_{\substack{0 \leq \alpha \leq l-3 \\ 0 \leq \beta \leq 2l-3}}.$$

Namely, this 1st hypercohomology group is generated by those $(l-1)(l-2)$ many elements over the ring $\hat{\mathbb{Z}}_p \otimes_{\mathbb{Z}} \mathbb{Q}$. Note that in the computation of the 1st hypercohomology, we have the cohomologous relation, i.e., $X^{i+l-2} Y^{\beta-l+1} dX$ is cohomologous to $X^{i+2l-2} Y^{\beta-l+1} dX$ over $\hat{\mathbb{Z}}_p \otimes_{\mathbb{Z}} \mathbb{Q}$. This is because the first spectral sequence

$$E_1^{p,q} = H^q(\underline{U}, \Omega^p_{\hat{\mathbb{Z}}_p}(\underline{U})\dagger \otimes_{\mathbb{Z}} \mathbb{Q})$$

vanishes for $q > 1$. Consequently, $\hat{\mathbb{Z}}_p \otimes_{\mathbb{Z}} \mathbb{Q}$-adic 1st cohomology $H^1(U, \hat{\mathbb{Z}}_p \otimes_{\mathbb{Z}} \mathbb{Q})$ is isomorphic to $Coker(O_U \dagger \otimes_{\mathbb{Z}} \mathbb{Q} \xrightarrow{d^0} \Omega^1_{\hat{\mathbb{Z}}_p}(U) \dagger \otimes_{\mathbb{Z}} \mathbb{Q})$, which is

$Coker((\hat{\mathbb{Z}}_p[X,Y]\dagger/(X^l+Y^l-1)) \xrightarrow{d^0} \Omega^1_{\hat{\mathbb{Z}}_p}(\hat{\mathbb{Z}}_p[X,Y]\dagger/(X^l+Y^l-1)))$.

Since $H^1(-, \hat{\mathbb{Z}}_p \otimes_{\mathbb{Z}} \mathbb{Q})$ is a contravariant functor from the category of non-singular affine varieties to the category of $\hat{\mathbb{Z}}_p \otimes_{\mathbb{Z}} \mathbb{Q}$-modules, a morphism f from $\underline{U}$ to $\underline{U}$ induces the endomorphism $H^1(f, \hat{\mathbb{Z}}_p \otimes_{\mathbb{Z}} \mathbb{Q})$ on $H^1(U, \hat{\mathbb{Z}}_p \otimes_{\mathbb{Z}} \mathbb{Q})$. If we define f in characteristic zero as follows, then the induced mapping becomes the p-th power mapping over the finite field F_p.

(*) $$\begin{cases} f(X) = X^p \\ f(Y) = Y^p \left(\sum_{k=0}^{\infty} \binom{1/l}{k} \left(\frac{-pT}{u} \right)^k \right) \end{cases}$$

where $u = (1-X^l)^p$, $-pT = 1 - X^{lp} - (1-X^l)^p$, and

$\binom{1/l}{k} = \frac{1}{l} \cdot \left(\frac{1}{l} - 1 \right) \cdot \dots \cdot \left(\frac{1}{l} - k + 1 \right)$. Then the endomorphism $H^1(f, \hat{\mathbb{Z}}_p \otimes_{\mathbb{Z}} \mathbb{Q})$ induces the $(l-1)(l-2)$ square matrix on the $\hat{\mathbb{Z}}_p \otimes_{\mathbb{Z}} \mathbb{Q}$-module $H^1(U, \hat{\mathbb{Z}}_p \otimes_{\mathbb{Z}} \mathbb{Q})$. Namely, by (*) we have

$$H^1(f, \hat{\mathbb{Z}}_p \otimes_{\mathbb{Z}} \mathbb{Q})(X^\alpha Y^{\beta-l+1} dX)$$
$$= X^{\alpha p} \cdot f(Y)^{\beta-l+1} d(X^p)$$
$$= pX^{\alpha p+p-1} \cdot Y^{p(\beta-l+1)} \cdot \left(\sum_{k=0}^{\infty} \binom{1/l}{k} \left(\frac{-pT}{u} \right)^k \right)^{\beta-l+1} dX.$$

Recall that the original Veneziano amplitude has the following integral representation:

$$A(s,t) = \int_0^1 x^{-\alpha(s)-1}(1-x)^{-\alpha(t)-1} dx$$

$$A(s,t) = \frac{\Gamma(-\alpha(s))\Gamma(-\alpha(t))}{\Gamma(-\alpha(s)-\alpha(t))}$$

where $\alpha(s) = \alpha_0 + \alpha' s$, $s = (k_1+k_2)^2$, $t = (k_1+k_3)^2$. Namely, the Veneziano amplitude is the convolution of two characters on the real axis. See [77] and [78] for details.

The p-adic analog of the Veneziano amplitude in terms of p-adic gamma function Γ_p becomes

$$A_p(s,t) = \frac{\Gamma_p(-\alpha(s))\Gamma_p(-\alpha(t))}{\Gamma_p(-\alpha(s)-\alpha(t))}.$$

Then the integral representation becomes the convolution of two multiplicative characters on $F_p^* = F_p - 0$:

$$\sum_{x \in F_p^*} \chi_a(x)\chi_b(1-x),$$

that is, the Jacob sum, denoted as $J(\chi_a, \chi_b)$. Then the Gross-Koblitz formula gives

$$-J(\chi_a, \chi_b) = \frac{\Gamma_p(a/l)\Gamma_p(b/l)}{\Gamma_p((a+b)/l)},$$

where $1 \leq a, b, a+b \leq l-1$. Then $J(\chi_a, \chi_b)$ can be computed as (-) trace of the square matrix induced by the endomorphism $H^1(f, \hat{\mathbb{Z}}_p \otimes_{\mathbb{Z}} \mathbb{Q})$ on $H^1(U, \hat{\mathbb{Z}}_p \otimes_{\mathbb{Z}} \mathbb{Q})^\rho$, where $\rho = \tilde{\chi}_a \times \tilde{\chi}_b$ such that $\tilde{\chi}_a(x) = \omega\left(x^{(p-1)/l}\right)^a$ where $\omega(x)$ is the Teichmuller representative of $x \in F_p^*$. See [47, especially Chapter III].

Prelude

We are ready to define the notion of a temporal topos, abbreviated as t-topos. Before the definition of a t-topos, we need categorical sheaf theoretic methods prepared in Chapters I and II which are capable of formulating quantum physical micro notions together with macro notions. The five axioms $(t-0)$ through $(t-4)$ for our structural aspect of our temporal topos theory provide both microcosm phenomena like entanglement, wave-particle duality, uncertainty principle, quantum fluctuation, light cones, dynamical aspect of space-time induced by particle mass, black holes, and the big bang type singularities. In our approach, the main devices for describing singularities in both microcosm and macrocosm levels are the notions of inverse and direct limits. See Definition I.1.6 in Chapter I for the notions of the inverse and direct limits. The notions of inverse and direct limits are defined categorically as universal objects satisfying universal mapping properties. The auxiliary notions in Chapter I of presheaves, the functor category, inverse and direct limits, and Yoneda's Lemma enable us to give such formulations in terms of categories and sheaves to both microcosm and macrocosm phenomena as mentioned above. The temporal topos theory provides both *background* and *scale independence* formulations of such micro and macro phenomena. Our approach is a categorical-structure dependence theory. In our formulation, the difference between the microcosm and the macrocosm, for example, depends upon the numbers of the factored morphisms of a morphism and the decomposed presheaves of a presheaf associated with a particle.

A history of exact science tells us that the task of capturing the truer natures of space and time in both micro and macro levels has been a perplexing problem. In t-topos, space-time is in a way given a treatment as particles. Namely, among the presheaves associated with particles, the presheaf associated with space-time plays the role of the terminal presheaf (object) in the category of presheaves. That is, for an arbitrary presheaf representing a particle, there is a unique morphism of functors from the presheaf to the presheaf representing space-time. When one measures space-time presheaf (over a generalized time period), one receives information of every presheaf (particle) via the unique morphism from the presheaf to the space-time associated presheaf. Our categorical-sheaf theoretic formulations with Yoneda's Lemma extended enable us to provide methods for formulating such interplays among space, time, and matter consistent for both microcosm and macrocosm. The emergence of a macro object from microcosm may be interpreted as the case where the presenting presheaf for an entity is also a sheaf.

Any theory adequate to describe the fundamental structures of the micro and macro universes (whether such crucial properties as Heisenberg's uncertainty principle, quantum entanglement, light cones, and some aspects of quantum cosmology can be derived from the theory or not) should be tested. Within the mental construct that created such a difficult problem as quantum gravity, it might be too difficult to find a satisfying solution. It might be better to begin with pure thoughts (pure mathematics) without being affected by our macrocosm common sense of three dimensional space plus one-dimensional time mentality to capture such bizarre and extraordinary notions as the Big Bang and the Planck scales of time and length.

Four fundamental principles of quantum physics are wave-particle duality, uncertainty, entanglement, and quantum fluctuations. All of these perplexing phenomena of microcosm are expressed as the interplay between presheaves and t-site objects. Within our t-topos methods, what will follow are consequences derived mainly from the following axioms $(t-0)$ through $(t-4)$. For the t-topos interpretation, the failure of determinism in quantum physical phenomena is a consequence of the non-uniqueness of a factorization of a t-site morphism associated with the canonical contravariant nature of a presheaf determining the next reified state. Furthermore, to be a theory truly leading to quantum gravity, every theory must be, as a necessary condition, a background independence theory. One must explicitly formulate the space-time effect by a particle mass even in the micro level as well as macro level within the theory, i.e., without changing the theories and their methods, for example, from quantum mechanics to general relativity. For instance, within one theory the microcosmic description of the gravitational effect of an electron needs to be formulated, as it should be done for cosmological objects whether the electron is in a particle state or a wave state. A t-topos theoretic formulation of such a microcosm effect is found in Remarks III. 1. 2 (3) and more generally in Remarks III. 5. 8 in the relativistic version. Such a theory must be scale-independent but rather it should be categorically structure-dependent in the following sense. The concept of a scale need not be determined by the assignment of a real number (or complex number, or even p-adic number) for the evaluation of physical quantity, but the description of scales of physical entities should come from the inner structures within the theory, for example, by the number of subobjects (subpresheaves) in a (micro-) decomposition of an object (a presheaf). Basically, the bigger that number is, the more macroscopic the corresponding object is. In our t-topos theory, the scale of time length whether microcosm or macrocosm is determined by the number of the factorized objects and linearly t-ordered morphisms of a morphism between two objects in the t-site. That is, the scale concept can be defined by the numbers in the categorical descriptions of the phenomenon, i.e., in this sense there are no distinctions from a microcosmic scale to a macrocosmic scale. Namely, microcosm and macrocosm are naturally induced from the categorical and sheaf theoretic structures in terms of t-topos theory. The main devices of t-topos are the following:

(i) The notion of a decomposition of a presheaf, especially the definition of a microdecomposition of a presheaf representing a particle.

(ii) The notion of a micromorphism in the t-site. The t-topos theoretic uncertainty principle is expressed as a consequence of a micromorphism in a t-site.

(iii) The concept of a linearly t-ordered morphism, corresponding to the classical notion of linearly ordered time is also a fundamental device.

(iv) The roles of inverse and direct limits for (ur-) singularities (defined by the universal mapping properties). This method also allows us to study the ultra-microcosm as the inverse (or direct) limits of certain sequences.

(v) The gravity sensitivity of a factoring of a morphism into linearly t-ordered morphisms, i.e., the temporal topos theoretic gravitational hypothesis, referred to as the t-g. hypothesis.

One might regard our temporal topos theory suited for categorical, structural, or conceptual aspects of quantum gravity.

Chapter III Temporal Topos

Section III. 1 Associated Presheaves and Space-Time sheaves; Background Independency

For the last eighty years, general relativity has been used for understanding the behaviors of astronomical scale objects and for studying the nature of black holes and the universe itself. Traditionally, differential geometry and related fields have been the main mathematical languages for general relativity theory. In particular, Riemannian geometry has provided a frame for general relativity. The original paper [65] by B. Riemann is worthy of reading. See also H. Weyl's [83]. On one hand, by capturing the gravitational field caused by matter, space-time changes from the flat Minkowski space-time to a curved four dimensional (pseudo-) Riemannian metric manifold, and on the other hand, quantum mechanics is for studying microscopic objects, e.g., nuclei and elementary particles where functional analysis has provided a frame, e.g., capturing a pure state as a unit vector in a Hilbert space and physical quantities as self-adjoint operators in the Hilbert space.

These two different frames have been applied for the same universe depending upon the scales, i.e., macrocosm or microcosm. We need to, or are even obliged to look for at least one consistent theory. Such a theory should be capable of determining the scales in terms of the structural consequences derived from such a theory, especially extremely near the big bang, namely, less than 10^{-43} seconds after the big bang, referred to as Planck time. Namely, our theory provides the following, for example: whether the time interval during two observed (particle) states is a quantum scale or not depends upon the number of the factored linearly t-ordered morphisms and objects of a given morphism between two objects in the t-site. For the definition of a linearly t-ordered morphism, see $(t-2)$ below, and see also Definition III. 3. 1. Namely, the usual distinction between micro and macro by assigned scales as real numbers can be replaced by the number of (micro-) factorizations of morphisms of the t-site and the number of the (micro-) decompositions of a presheaf in the t-topos. See the following Definition III.3.1 and Definition III.3.2. As we will define, a morphism without any such factorization is used to describe phenomena in the (ultra-) microcosm. In terms of a particle view, when a presheaf associated with a particle can be decomposed into non-trivially (non-isomorphic) subpresheaves, we consider such a particle is in a macrocosm scale. On the other hand, when a presheaf cannot be decomposed into non-isomorphic subpresheaves, we consider such a presheaf corresponds to an elementary particle. One can define such definitions for a morphism and an object in the t-site which are treated in Sections 3 and 4 of Chapter III.

One of the main reasons for introducing categorical and sheaf theoretic methods to quantum field theory, for example, is to avoid divergent expressions, e.g., for the total amplitude of a quantum process. We would like to mention the possibility that our sheaf theoretic method is relevant to some non-perturbative approaches to quantum gravity, e.g., loop quantum gravity and non-perturbative superstring theory. We are offering one of the unifying frames for both the micro and macro worlds as described briefly above. In this model, called temporal topos, for example, when one physical quality (e.g., a position of a particle) of a particle is measured, one category is assigned for the measurement of the particle. Hence, we need to introduce the notion of a product of categories for our theory. For a particle, we associate a presheaf m. A state of the presheaf m is controlled by an object of the domain category called a t-site. In our theory, however, m need not be defined for every object of a site. (See Definition 3.1, Chapter I, for the notion of a site.) One could rename such an m as a pseudo-presheaf (or a contravariant quasi-functor), since the usual notion of a presheaf is defined for all objects of a site. However, we call such a restricted presheaf simply a presheaf in this treatise. As we will define in the following $(t-1)$, when $m(V)$ is defined, we say m is in an ur-particle state over V. As we shall see, for a linearly t-ordered *micromorphism* $V \to V'$ (for this notion, see the following Section III. 3), there does not exist any ur-particle state between the two linearly ordered states corresponding to V and V', which is relevant to the phenomenon quantum tunneling. The impossibility of any factorization of a micromorphism implies that it is impossible to locate the position-time of the corresponding particle associated with the presheaf m. Hence, it is impossible to observe the particle between the time interval corresponding to V and V'. Therefore, it is a meaningless question to ask about the location of the particle between the two ur-particle states corresponding to V and V'. That is, the notion of position itself is not definable for that time interval. Namely, the state of the presheaf m between V and V' must be in an ur-wave state. Note also that as we mentioned above, this ur-wave state period is the period where quantum tunneling not only can, but also must occur under a certain restriction. We will come back to the topic of quantum tunneling in Section III. 3. Note that for our theory of t-topos, whether a quantum tunneling can occur or not is a t-site object dependent notion.

Definition III. 1.1 Let C_β be a category indexed by a finite set Λ. Consider the product category $\prod_{\beta \in \Lambda} C_\beta$, as defined in Example I. 1.3 (5). The category $\hat{S}$ of (contravariant) functors from a site S (See Definition I. 3.1, i.e., a site is a category with a Grothendieck topology whose objects are said to be *generalized time periods* when used in our more restricted sense.) to $\prod_{\beta \in \Lambda} C_\beta$ is said to be a *temporal topos* (which is often abbreviated as *t-topos*) when the following axioms from $(t-0)$ through $(t-4)$ are satisfied.

$(t-0)$: As we have already introduced in Chapter I, let κ, τ, and ω be space, time, and space-time sheaves, respectively. Namely, κ, τ, and ω are not only presheaves but also assumed to be sheaves, (i.e., Definition I. 3. 2.). Furthermore, space-time sheaf ω is a *terminal object* of $\hat{S}$. (Note that all the terminal objects are isomorphic, hence sheaf ω may be said to be *the* terminal object of $\hat{S}$.) Namely, for any object m of $\hat{S}$, there exists a unique morphism $\sigma_m : m \longrightarrow \omega$ of functors. Generally speaking, a terminal object in the category of presheaves is a constant functor assigning a terminal object for any object in the t-site. Namely, later we will be describing explicitly the mass effect on the space-time sheaf in Remark III. 1. 2, and in Section III. 5 as the t-g. hypothesis. Whenever space-time sheaf ω is observed, one is also measuring the effects of presheaves in $\hat{S}$ by the compositions of morphisms of $\sigma_m : m \to \omega$ for all $m \in \hat{S}$ and the observation morphism from ω to an observer over a generalized time period within the light cone in the sense of the t-topos. See the following $(t-1)$ for the physical meaning of a presheaf representing a particle. Also for the t-topos theoretic notion of a light cone, see Section III. 5. We sometimes use the notation ω_m or even $\omega(m)$ for the space-time sheaf associated with an arbitrary m in $\hat{S}$ in our previous papers. Note that $\sigma_m : m \to \omega$ can be written as $\tilde{\omega}(m)$, where as morphisms in $\hat{S}$ we have
$\tilde{\omega}(m) = Hom(m, \omega) = Hom(-, \omega)(m)$. See Section I. 2 of Chapter I for this notation and its meaning. For this notational motivation, see Remarks I. 2. 2 (4). Then ω can be regarded as a contravariant functor on the category $\hat{S}$ of presheaves on the t-site S via the Yoneda Embedding Principle, referred to as (Y.E.P.) in Remark I. 2. 2 (4). Capturing ω as the terminal object of t-topos $\hat{S}$ corresponds to the diffeomorphism invariance notion in the sense that locally defined space-time sheaf ω (i.e., defined over objects of the t-site) depends upon a morphism σ_m from each presheaf m of $\hat{S}$ representing a particle. See also the following Remarks III. 1. 2 (3).

A t-topos without such a terminal object might be considered as a universe without space-time, i.e., no universe. As for identity, we assume that all the presheaves representing particles mutually are distinct but can be isomorphic. Namely, let for example, γ and γ' represent photons. Then we have an isomorphism $\gamma \xrightarrow{\approx} \gamma'$ but $\gamma \neq \gamma'$.

$(t-1)$: For any particle $\bar{m}$, there always exists a presheaf m in $\hat{S}$. However, such a particle presenting presheaf m, $m(V)$ need not be defined for an arbitrary object V of the t-site S. Then the presheaf m is said to be *associated with* the particle $\bar{m}$, or m is *representing* the particle $\bar{m}$. When $m(V)$ is defined for an object

V of S, the presheaf m is said to be in an *ur-particle state (or particle ur-state)* over (or during) the generalized time period V. When an object V in S is not specified so that $m(V)$ may be defined, the presheaf m is said to be in an *ur-wave state* (or *wave ur-state*) or an *ur-superposition state*. We also say that presheaf m is *reified* (or *compatible*) at V when $m(V)$ is defined for an object V in S. That is, a non-reified m of the t-topos $\hat{S}$ is by definition in an ur-wave state. When a presheaf m is in an ur-particle state, one could measure the mass of the particle, for example, represented by the presheaf m via the projection from the product category to a particular category. One can say that the collapse of the wave state of a particle in the classical sense corresponds to the unique choice of an object of the t-site S over which the presheaf representing a particle is reified. Note that our ur-superposition expression of a presheaf $m(-)$, whose notation is often used in category theory, without the evaluation at any object of the t-site may be considered to correspond to the usual notation of the state vector $|\psi\rangle$.

($t-2$): Let m be a presheaf representing a particle as in ($t-1$). Suppose that the particle is twice observed with V and U being the objects of the t-site determining these corresponding ur-particle states of m. Let $V \xrightarrow{g} U$ be the corresponding morphism in S so that $\tau_m(V)$ precedes $\tau_m(U)$ in the usual linearly ordered time sense corresponding to the first observation at V and the second observation at U of the particle, respectively. Then such a morphism $V \xrightarrow{g} U$ is said to be a *linearly t-ordered morphism (or t-linearly ordered morphism)* from V to U. We postulate that the t-site S *has enough objects* which means the following. For an arbitrary sequence $---\xrightarrow{g_{i-1}} V_i \xrightarrow{g_i} V_{i+1} \xrightarrow{g_{i+1}} ---$ of linearly t-ordered morphisms and objects of t-site, $V_k \neq V_l$ must hold for $k \neq l$. We should allow isomorphic objects for $k \neq j$.

It is plain in our formulation, but not trivial to assert the following: for a linearly t-ordered isomorphism $V \xrightarrow{g} U$, since m is functor, the reified ur-particle states corresponding to V and U need to be isomorphic, however $m(V) \neq m(U)$. The presheaf m itself is invariant, but there corresponds a new ur-state object $m(V)$ for each object V of the t-site. Namely, even though two experiments for one particle presented by a presheaf m, corresponding to a linearly t-ordered morphism, are done under the same condition, in general the corresponding (ur-) states differ. That is, on a particular occasion in the t-topos sense means that a particular object V of the t-site is assigned. This is the t-topos aspect concerning the measurement problem. Our t-topos formulation implies that there is no such thing as being under the same condition for two experiments done at two different times.

Note that for example, for a covering $\{U \xleftarrow{g_i} U_i\}$ of U in Section III. 4, morphism g_i need not be linearly t-ordered. See Section III.5 for the connection between our formulation of a light cone and a linearly t-ordered morphism. It is fundamental for our theory that for a reified ur-particle state $m(V)$ of m with an object V of the t-site, more than one linearly t-ordered morphism may exist from V to another object of the t-site. Note also that for a linearly t-ordered morphism $V \to U$, there may exist other factorizations which need not be by linearly t-ordered morphisms of $V \to U$. As we shall see in what will follow, Heisenberg uncertainty, the notion of a complex superposition consisting of alternative histories, and quantum fluctuations together with the notion of a microdecomposition in III. 4 are relevant to above formulations in terms of t-topos.

$(t-3)$: An *observation (measurement)* of m by P of $\hat{S}$ over a generalized time period V of the t-site S is defined as a morphism of functors over V. Or one can phrase this notion as follows. For objects m and P in $\hat{S}$, an observation (measurement) from m to P during the generalized time period V is a natural transformation s defined over V. Namely, we have a morphism of functors $s_V : m(V) \longrightarrow P(V)$ evaluated at V. Notice that the only difference between the observer and the observed in the t-topos approach is the direction of a morphism between them. Consequently, in the dual (opposite) category, the role of the observed and the observer is reversed. Namely, in the dual category $(\prod C_\beta)^{opp}$ we get the observation (measurement) natural transformation over V as $s_V^{opp} : P(V) \longrightarrow m(V)$. Beyond this formal consequence from the duality, one may consider physical applications of the duality.

Note that the corresponding relativistic version of the definition of the above notion of a measurement within a light cone in macrocosm (or in microcosm) will be given later in Section III. 5. The above formulation of an observation in $(t-3)$ is valid only when m is "close enough" (i.e., non-relativistic case) to an observer in the sense that the associated time with observed $\tau_m(V)$ and the associated time with the observer $\tau_P(V)$ are nearly simultaneous with respect to the t-site S.

Consequently, for a presheaf m associated with a particle to be measured (or observed), m must be in an ur-particle state. Namely, it is a necessary condition to be in an ur-particle state for m to be observed. In the jargon of the subject, this abrupt change of the state from being in an ur-wave state to being in an ur-particle state corresponds to the expression "collapse of the wavepacket" in quantum physics. Note the necessary condition (not a sufficient condition) indicates that a presheaf m need not be in an ur-wave state when m is not observed. That is, m may be in an ur-particle state even when m is not observed. The dynamical development in terms of t-topos between two observations corresponding to a linearly t-ordered morphism $V \xrightarrow{g} V'$ begins with the ur-particle state at V. Then m is in the ur-wave state until *collapsing* onto the ur-particle state again at V'.

A plain, but important remark is the following. For a particle associated presheaf m to be measured (or observed) by a presheaf P, not only m but also P need to be in ur-particle states. One can also say that the specification of an object of the t-site to be in an ur-particle state may be regarded as a disturbance to the ur-wave state.

Axiom $(t-4)$ for a covering will be given in III. 4.

Remarks III. 1. 2 (0) Like any other approaches to the foundation of quantum mechanics in terms of (pre-) sheaves (topos), physical states vary discretely, i.e., in non-continuum. Temporally and position-wise nearest observable states are provided by the concept of micromorphisms. (For the definition of a micromorphism, see Definition III. 3.1.) Recall that $(t-1)$ says: for every particle there corresponds a presheaf. Namely, we postulate $(t-1')$: for an arbitrary presheaf, there may not correspond any particle. We treat both space-time and particles as represented by presheaves. In this sense, our temporal topos theory is a *presheaf gravity*. That is, we presheafify not only particles but also space-time in the sense of representing the physical entities by presheaves. (Notice that this presheafification process is not in the sense of *sheafification of a presheaf* as constructed and characterized as a sheafification functor in sheaf theory, and that our notion of "representing entities by presheaves" does not mean categorically represented by presheaves as in Definition I. 1. 12.) We will provide the consequences in what will follow from our theory for microcosm and macrocosm based on the sheaf-category methods.

(1) For example, the *wave-particle duality* of an electron can be phrased in our theory as the difference between the two ur-states of a presheaf, i.e., whether the presheaf e associated with the electron is reified or not. Namely, a presheaf e is in an ur-particle state when an object V in the t-site is both *specified and reified* as $e(V)$. Then presheaf e is in an ur-particle state as an object $e(V)$ of the product category $\prod_{\beta \in \Lambda} C_\beta$.

Let $V \xrightarrow{g} V'$ be a linearly t-ordered morphism where a presheaf m is in ur-particle states at V and V'. Let us consider the totality

$$\{V \xrightarrow{g^i_{0,1}} V_{i_1} \xrightarrow{g^i_{1,2}} V_{i_2} \to ---- \to V_{i_{N(i)}} \to V'\}, i \in I.$$

Namely, we consider that there are $|I|$ (micro-) factorizations of $V \xrightarrow{g} V'$, where $|I|$ is the finite cardinality of the index set I, and for the i-th factorization there are $N(i)$-many t-site objects between $V \xrightarrow{g} V'$. That is, morphism g is factored by $N(i)$ morphisms for each $i \in I$. Note that some of the morphisms $\{g^i_{j,j+1}\}$ appearing in the factorization may not be linearly t-ordered. One can consider the totality

$$\{V \xrightarrow{g^i_{0,1}} V_{i_1} \xrightarrow{g^i_{1,2}} V_{i_2} \to ---- \to V_{i_{N(i)}} \to V'\} \text{ of all the possible factorizations of}$$

$V \xrightarrow{g} V'$ as the corresponding notion to the configuration space of Feynman. We will come back to this topic as we introduce further t-topos theoretic notions relevant to this issue.

(2) In t-topos, as the terminal object of $\hat{S}$, space-time sheaf $\omega = (\kappa, \tau)$ is a special presheaf in the following sense. For every particle there exists a presheaf describing the various states of the particle including the ur-wave state. Every presheaf need not be associated with a particle, i.e., $(t-1')$. However, space-time

sheaf ω belongs to the same category, i.e., the t-topos of presheaves. That is, particles and space-time are objects of the t-topos as mentioned earlier. If the unique morphism $\sigma_m : m \to \omega$ in $(t-0)$ is a monomorphism, one might consider ω as associated with a largest particle. Note that space-time sheaf ω evolves with the states of particles associated presheaves. Namely, a (non-relativistic) measurement of space-time ω during a generalized time period V of the t-site by P is associated with the mass effects on space-time via the morphisms $m \xrightarrow{\sigma_m} \omega \xrightarrow{s} P$ evaluated over V. In this sense, t-topos is a background independence theory, namely, space-time is defined and measured locally over V. The *sheaf theoretic position (space-time) measurement* by P of a particle represented by the presheaf m can be defined by the composition morphism $s \circ \sigma_m$:

$$m \xrightarrow{s \circ \sigma_m} P$$
$$\searrow \quad \nearrow$$
$$\omega.$$

It makes no sense to make a comment on a position of a particle unless the associated presheaf is reified over an object of the t-site. Namely, for an ur-wave state presheaf (particle), a notion of the position itself does not exist. We will make further explicit comments on this issue and relevant consequences in what will follow after appropriate concepts have been defined. Our approach says that space-time as the terminal object of $\hat{C}$ is not only dependent upon presheaves associated with particles but also on an observation of space-time made by an observer over a generalized time period of the t-site. Namely, one can not measure space-time ω without particle presheaves and together with specified objects of t-site. That is, our formulation above implies the following. When one measures the space-time presheaf ω over a generalized time period V, all the effects caused by all the presheaves on ω e.g., by masses, are measured by the compositions of morphisms from presheaves in the light cone with an observation morphism $\omega(V) \xrightarrow{s_V} P(V)$ from ω to the observer P over V as in $(t-3)$. See (3) of this Remark. See also Remark I. 2. 2. As mentioned earlier, we emphasize this nature by denoting the time, space, and space-time associated with presheaf m also by τ_m, κ_m and ω_m, respectively. Also note that the classical notion of vacuum may be best interpreted in terms of space-time sheaf ω. As a terminal object of the t-topos, ω alone exists; however, as long as particles with mass exist, one cannot measure only ω. Namely, when one measures ω, one must measure all the effects via the compositions of the canonical morphisms $\{\sigma_m\}_{m \in Ob(\hat{S})}$ in the t-topos theoretic light cones. (See III. 5.)

(3) We will make several technical remarks on space-time presheaf. By Remarks I. 2. 2. (2), for the space-time sheaf ω_m in the above (t-0), we have

$$\omega_m(V) = (m \xrightarrow{\sigma_m} \omega)(V) = m(V) \xrightarrow{(\sigma_m)_V} \omega(V). \qquad \text{(III.1.1)}$$

This morphism says, in light of $(t-3)$ of Definition III. 1. 1., that space-time presheaf ω (as a terminal object of $\hat{S}$) can be regarded as a *universal measurement presheaf* for all the presheaves $\{m\}$ of $\hat{S}$, as we indicated in (2) above. Namely, by measuring the space-time sheaf ω over V, we get the measurements of the states of all the presheaves m over V, in the t-topos theoretic light cone in Section III. 5., via the morphisms $(\sigma_m)_V$. For example, space-time sheaf is the object of $\hat{S}$ where the relation between the curvature of space-time and the energy-matter may be recovered in terms of composition of the morphism σ_m and an observation morphism s_V below together with the t-g. hypothesis in Section III. 5. As we shall see later, the above formulation is for the non-relativistic case, i.e., such a morphism as in (III. 1. 1) can exist only when both m and ω are close enough, or more generally in the relativistic terms: in a *mutually intersecting universal light cone*. See Definition III. 5. 4 for the general case. Note that when one considers such a notion as a light cone in microcosm, we need the relativistic formulation since the space-time can only be observed discretely.

As we saw in Chapter I, one may be able to carry out formulating a t-topos theory by always identifying ω as the embedded object $\tilde{\omega} = Hom_{\hat{S}}(-,\omega)$. That is to say, the space-time sheaf ω may be (and can be) considered as a contravariant functor $\tilde{\omega} = Hom_{\hat{S}}(-,\omega)$ from the t-topos $\hat{S}$ to the product category $\prod_{\beta \in \Lambda} C_\beta$. In terms of Little Zen of Yoneda in (4) of Remarks I. 2.2, i.e., identifying $\tilde{\omega} = Hom_{\hat{S}}(-,\omega)$ with ω, these formulations are done at the same level (under the assumption that an object of the codomain category is at least a set). See also the diagram of functors in Section I. 3.

Furthermore, we need to consider the effect of an observer itself in our formulation. First recall that a measurement by an observer of the space-time effect caused by the particle with mass can be formulated as follows. When an observer P measures the space-time effect by the presheaf m associated with a particle over a generalized time period V of the t-site S, the non-relativistic version of the diagram should be expressed as

$$m(V) \xrightarrow{(\sigma_m)_V} \omega(V) \xrightarrow{s_V} P(V) \qquad (\text{III.1.2})$$

where s_V can be regarded as an observation morphism from the space-time sheaf ω to P over V, and $(\sigma_m)_V$ is the canonically induced morphism of functors (natural transformation) in $(t-0)$ evaluated at V. We will give the relativistic version of this formulation in Section III. 5 where all the generalized time periods in the above diagram may differ, but they belong to a mutually intersecting light cone. Consequently, space-time ω cannot be observed (measured) without a particle effect caused by a presheaf m, since ω is a terminal object of $\hat{S}$. The image of the composition morphism

$$S_V \circ (\sigma_m)_V : m(V) \xrightarrow{S_V \circ (\sigma_m)_V} P(V) \qquad (\text{III.1.3})$$

is the information of m obtained by the observation via space-time ω over V. In a projected category from the product category $\prod_{\beta \in \Lambda} C_\beta$, the image of the morphism in (III. 1. 3) may be considered as the measurement of the field caused by the particle presented by presheaf m via space-time sheaf ω. Next consider the case where two entities are involved. Let m and m' be presheaves representing two particles. As a non-relativistic case consider the following diagram.

$$m(V) \xrightarrow{(\sigma_m)_V} \omega(V) \xleftarrow{(\sigma_{m'})_V} m'(V)$$
$$\downarrow$$
$$P(V). \qquad (\text{III. 1. 4. } m')$$

Then for $S_V : \omega(V) \longrightarrow P(V)$, the compositions $S_V \circ (\sigma_m)_V$ and $S_V \circ (\sigma_{m'})_V$ give the measurement on both m and m' as the image of the induced compositions $S_V \circ (\sigma_m)_V$ and $S_V \circ (\sigma_{m'})_V$. As long as one needs to formulate an effect in microcosm from mass on the space-time, e.g., from an electron and a proton, such a measurement formulation for ur-particle states as above is needed (even though the mass of an electron may be a half a thousandth of the mass of a proton). However, this formulation is not complete especially for microcosmic situation, since the measurement device P could interfere with the space-time ω as P measures m. In the microcosm, we need to replace the above (III.1.2) with a diagram expressing the effect from P as well. Namely, we need a formulation considering the effect on the space-time sheaf, as a morphism from P to ω, caused by a measuring device (an observer) P itself in $\hat{S}$. That is, we replace the above m' with P in diagram (III. 1. 4. m'). Diagrams (III.1.2) and (III.2.3) would be sufficient, at least as an approximation, for the case when the observer is a distance away from the observed m as in the macro case. Secondly, we will consider the effect of the observer on the space-time as follows. (However, the truly genuinely relativistic case of those diagrams for both microcosm and macrocosm where the above single object V in the t-site needs to be replaced by certain t-site objects U and W in the appropriate light cones will be given in III.5.) As mentioned above, we need to examine the following diagram, especially in a microscopic case, where the interference of the observer itself with space-time can not be ignored. First, since ω is a terminal object in the category $\hat{S}$, we have from Axiom $(t-0)$

$$m \xrightarrow{\sigma_m} \omega \xleftarrow{\sigma_P} P$$
$$\downarrow^{S} \quad \nearrow_{id.} \qquad (\text{III.1.4. } P)$$
$$P$$

or, as the diagram evaluated at V, we have:

$$m(V) \xrightarrow{(\sigma_m)_V} \omega(V) \xleftarrow{(\sigma_P)_V} P(V)$$
$$\downarrow{s_V} \quad \nearrow_{(id.)_V} \quad \quad \text{(III.1.5)}$$
$$P(V)$$

Namely, when P measures space-time ω over V via the observation morphism S_V, P receives the measurements both $S_V \circ (\sigma_m)_V$ on m and $S_V \circ (\sigma_P)_V$ on the observer P itself over the generalized time period V. That is, since space-time presheaf is the terminal object of $\hat{S}$, the observer dependency formulation of the space-time effect (not only caused by the observed but also by the observer) is a natural consequence. One can consider this formulation of space-time ω as a terminal object of t-topos $\hat{S}$ as the *background independence* notion in the sense of being locally defined on a t-site object. Namely, we have a notion of space-time not constantly existing behind presheaves representing particles as a background stage. Rather, space-time is dynamically evolving via morphisms $\{s \circ \sigma_m\}_{m \in Ob(\hat{S})}$ not only with particles but also with generalized time periods determining the states of the particles. As mentioned earlier, when space-time is measured over a generalized time period V by an observer P, as in (III. 1. 5), one can not measure space-time alone; one needs to consider all the morphisms from presheaves in the *mutually intersecting universal light cone*. The above formulation captures space-time ω as a dynamically evolving presheaf with varying states determined by objects (generalized time periods) of the t-site and the effects depending, via morphisms, upon all the objects of the t-topos $\hat{S}$ reified with objects of the t-site. For another crucial dynamical notion (i.e., the general relativistic aspect), see the t-topos theoretic gravitational hypothesis (referred to as the *t-g. hypothesis*) in Section III. 5 on black holes.

Next, consider the case where the above m and P are presheaves corresponding to microscopic objects. There may not exist a common t-site object V over which a measurement can be made.

(4) Let a presheaf m associated with a particle be observed twice over generalized time periods V and U. Suppose m is observed over V first and then over U, as in $(t-2)$. That is, time $\tau_m(V)$ precedes time $\tau_m(U)$ in the usual classical linearly ordered sense. Then the morphism g from V to U in the t-site S is a linearly t-ordered morphism as in $(t-2)$. Suppose that m is measured (or observed) by P over V. Namely, there exists a morphism s_V from $m(V)$ to $P(V)$, which is the definition of an observation (or measurement), i.e., $(t-3)$. Then we have the following diagram:

$$m(V) \xleftarrow{m(g)} m(U)$$
$$\downarrow \quad \swarrow_{s_V \circ m(g)} \quad \quad \text{(III. 1.6)}$$
$$P(V)$$

where the composition $s_V \circ m(g)$ in the above diagram should be understood as the measurement (information) of the future ur-particle state $m(U)$ over U by measuring $m(V)$ by $P(V)$. Namely, the image of the composite morphism $s_V \circ m(g)$ is the amount of information P can obtain about the ur-particle state $m(U)$ by measuring the state of m over V, preceding the state of m at U. Notice also that our t-topos says the following. When the ur-particle state $m(U)$ is observed by $P(U)$, we have no information on the preceding ur-particle state $m(V)$. This is because the observation morphism $s_U : m(U) \to P(U)$ cannot be composed with the canonical morphism $m(g) : m(U) \to m(V)$ in diagram (III.1.6) above. Namely, for example, when an electron is observed (over a generalized time period), one has no information about the earlier state of the electron.

(5) Without the evaluation over an object of the t-site, i.e., at the natural transformation (namely, morphism of functors) level, we do not yet know a physical interpretation of the unique morphism $m \xrightarrow{\sigma_m} \omega$ to the terminal object ω per se. For example, suppose that space-time was measure by P over a generalized time period U. Let us denote that measurement morphism of the space-time by s_U. Then the information P receives is the totality of compositions $\{s_U \circ \sigma_m\}_{m \in \hat{S}}$ from all presheaves within universal light cones. (See Definition III.5.2 for the definition of a universal light cone.) Then one needs to ask the effect, if it exists, from presheaf associated physical entities even outside the universal light cones with respect to U. We do not know an answer to this question since we do not know how to interpret the physical effects of the unique morphism $m \xrightarrow{\sigma_m} \omega$ without evaluations at objects of the t-site.

(6) Notice that the well known Exclusion Principle of Wolfgang Pauli can be rephrased in terms of t-topos as follows.

(EP) Let e and e' be electron-associated presheaves. Then there is an isomorphism $e(V) \xrightarrow{\approx} e'(V)$ if and only if we have the isomorphism $e \xrightarrow{\approx} e'$ of functors in $\hat{S}$.

The above principle should be read as: when a presheaf e associated with an electron has an ur-particle state, it is impossible for another electron presheaf e' to have the same (isomorphic) state unless e and e' are isomorphic.

(7) First let us consider a macro case. That is, for a presheaf m, a linearly t-ordered morphism $V \xrightarrow{g} U$, over which m is observed, has ample factorizations. As we noted, a measurement of the ur-particle state $m(V)$ by $P(V)$ gives the information of the future ur-particle state $m(U)$ by the composition of the canonically induced morphisms from $V \xrightarrow{g} U$ and the observation morphism from $m(V)$ to $P(V)$. Note, however, in the case of an unfactorable morphism, i.e., a micromorphism, with domain object W, there need not be a uniquely determined micromorphism leaving W (See III.3 for the definition of a micromorphism.). If it

were so, it would be deterministic in microcosm, which is relevant to the non-uniqueness of a factorization of such a macrocosmic morphism $V \xrightarrow{g} U$ considered above.

(8) Note also that in the dual category $\hat{S}^{opp}$ of the t-topos $\hat{S}$, space-time sheaf ω becomes an initial object of $\hat{S}^{opp}$. If P observes m in the dual category $\hat{S}^{opp}$, P receives information of the space-time ω by the composition $\omega \longrightarrow m \longrightarrow P$ of the morphisms. Notice that the action of observation in $\hat{S}^{opp}$ is not canonically (not functorially) induced simply by reversing the direction of a morphism. An observation is not a natural notion in the sense that we need to redefine an observation morphism of $\hat{S}^{opp}$ as an information flow, as done in $\hat{S}$.

(9) Even though space-time sheaf needs to be reified for measuring the space-time affected by presheaves expressed as the unique morphisms $\{\sigma_m\}_{m \in Ob(\hat{S})}$, the physical role in considering the notion of an ur-wave or ur-particle state of the space-time sheaf itself has not been given. This is simply because we do not know the role of presheaf ω, except as the terminal object to measure the effect of particles (presheaves).

Note III. 1. 3 We give categorical and technical remarks in our formulations and some consequences. Yoneda's Lemma in Section I. 2 enables us to consider directly $\omega(m)$ or $\tilde{\omega}(m)$ where $\tilde{\omega} = Hom_{\hat{S}}(-, \omega)$. That is, the element of the set $\omega(m)$, is the unique morphism of presheaves (as discussed in Definition I. 1. 9.) $m \xrightarrow{\sigma_m} \omega$. Regard ω as $\tilde{\omega}$, i.e., as a functor from $\hat{S}$ to the category of sets. For a family of morphisms $\{m \longleftarrow m_j\}_{j \in J}$ for the presheaf m, regarded as a covering of m in $\hat{S}$, consider the exact sequence

$$Hom_{\hat{S}}(m, \omega) \to \prod Hom_{\hat{S}}(m_i, \omega) \rightrightarrows \prod Hom_{\hat{S}}(m_i \times m_j, \omega).$$

Then the space-time ω is a sheaf for $\{m \longleftarrow m_j\}_{j \in J}$. In this sense, space-time sheaf ω can be considered different from presheaves associated with particles. That is, ω (or the representing presheaf $\tilde{\omega}$ as an object of $\hat{S}$) is a sheaf also over the t-topos $\hat{S}$. See also Remark III. 5. 8. Notice that these mathematical observations are induced from Little Zen of Yoneda's Lemma and Yoneda's Embedding. A project could be undertaken to find the physical interpretations of the above remark.

Epitome III.1 Having described the fundamental approach of our theory, we quote the following from well known *"On the Hypotheses Which Lie at the Bases of Geometry,"* by Bernhard Riemann [65], translated by William Kingdon Clifford, Nature, Vol. VIII, Nos. 183, 184, pp. 14-17, 36, 37. (Transcribed by D. R. Wilkins, Preliminary Version, December 1998) as follows.

"Section 3. The questions about the infinitely great are for the interpretation of nature useless questions. But this is not the case with the questions about the infinitely small. It is upon the exactness with which we follow phenomena into the infinitely

small that our knowledge of the causal relations essentially depends. ---- we seek to discover the causal relations by following the phenomena into great minuteness, ----

If we suppose that bodies exist independently of position, the curvature is everywhere constant, and it then results from astronomical measurements that it cannot be different from zero; or at any rate its reciprocal must be an area in comparison with which the range of our telescopes may be neglected. But if this independence of bodies from position does not exist, we cannot draw conclusions from metric relations of the great, to those of the infinitely small; ----- the empirical notions on which the metrical determinations of space are founded, the notion of a solid body and of a ray of light, cease to be valid for the infinitely small. We therefore are quite at liberty to suppose that the metric relations of space in the infinitely small do not conform to the hypotheses of geometry; and we ought in fact suppose it, if we can thereby obtain a simpler explanation of phenomena.

---- the reality which underlies space must form a discrete manifoldness, or we must seek the ground of its metric relations outside it, ----"

Generally, it is rather difficult to understand and learn from such a historical document as the above quoted *"On the Hypotheses Which Lie at the Bases of Geometry,"* by Bernhard Riemann; however, it is not quite meaningless to approach classical works with a newly developed theory seeking a certain degree of agreement or consistency with historically great works. It is our hope that capturing the concept of space-time as a (the) terminal object in the category of presheaves is supported (or at least non-contradictory to his view of space and time) by Riemann's statements quoted above. It was about ninety years ago when quantum theory was developed explaining non-common sense phenomena of amicrocosm. Namely, it became clear that for example, an electron is governed by totally different rules, which is far from our macro world physics. First of all, we should mention the Schrodinger equation. One of the purposes of quantum mechanics is to capture an electron as a wave so that we can describe movement and energy. Secondly, it is the probabilistic nature, e.g., to determine the position of an electron. This portion of microcosm nature corresponds to $(t-1)$ of Definition III. 1.1. That is, for a presheaf m associated with a microscopic object and for an arbitrary object V of the t-site S, $m(V)$ need not be defined. It is well known that there were disagreements among the founders of quantum theory on how to interpret the probabilistic nature of the Schrodinger equation. It was trusted historically that the full knowledge of a state in the past determines the unique state of a future state. This deterministic oriented view in terms of t-topos would be the following. Let $V \rightarrow U$ be a linearly t-ordered morphism in the t-site S over which a particle associated presheaf m is observed by P, and let $m(V) \rightarrow P(V)$ be a measurement morphism. However, as noted earlier, a factorization of $V \rightarrow U$ is not uniquely determined. A well-known case of this situation is the double slit experiment where V is the specified object of t-site corresponds to the state, for example, where an electron was fired, and U corresponds to the state hitting a screen. The morphism $m(V) \rightarrow P(V)$ does not determine which factorization of $V \rightarrow U$ needs to be chosen uniquely. That is, one cannot know which slit an electron must go through even if one has the full knowledge of the state $m(V)$ when it was fired. The best information one can anticipate is to know which one is more likely to happen. However, actuality need not happen at the likely place. As noted and defined in Definition III. 1.1. ($t-3$), an observation of a particle requires an object of the t-site. A question one can ask is whether when the particle is not observed, the particle is reified or not, i.e., whether there is an object W to obtain the reified $m(W)$ or not. The t-topos theoretic answer is affirmative for this question. Namely, when not observed, m may be defined for an object W in S. However, when m is observed over a generalized time period V by P, m is not only in the ur-particle state $m(V)$, but also there exists a morphism $m(V) \to P(V)$. (See Chapter III. 5 for the relativistic version of this remark.) Then P also observes the effect on m from P as the compositions of $s \circ \sigma_m$ and $s \circ \sigma_P$ in the diagram (III.1.5) over V. This is the formulation by t-topos theory to capture the reified states (i.e., ur-particle states) corresponding to when m is observed and when m is not observed. Note also that one could obtain the information of the ur-particle state $m(V)$ without the influence from P if one could delete $s \circ \sigma_P$ from the compositions.

One of the aspects of the t-topos theory, which has not been established, is the quantitative aspect. We have mentioned in our series of papers [39], [40], [41], [43], FUNC (Functional Composition Principle) of Isham (see also [30]) should play an important role in a quantitative study. The t-topos theoretic FUNC is a future project; however, we are to replace the field $\mathbb{R}$ of real numbers in FUNC possibly with a complete non-Archimedean field. See [9], [47], or the work of I. V. Volovich, e.g., [78]. There are two motivations behind this direction of our investigation on the replacement of $\mathbb{R}$. Not only the measured values for quantum quantities are discrete, but also at the sub-Planck level the value for a sum of two entities x and y should equal the larger of the two; it is not less than or equal to the sum of the two values of the entities x and y. Namely, the measurement needs to be non-Archimedean. See [77], [78], [76], and [9]. Before closing the first Epitome, brief philosophical and ontologically oriented comments may be appropriate. The issue is "What are the most fundamental entities of the universe?" It is said that Pythagoras considered the universe (possibly symbolically) as consisting of numbers. At the sub-Planck level, it may be not too certain that quarks, strings, and quantum fields are the truly fundamental entities. If one says presheaves are fundamental entities, it would be closer to Pythagoras' thought. Choosing presheaves (a topos) as describing objects as fundamental entities may not be considered as outrageous as choosing numbers as fundamental entities. If a single theory does not exist as a complete theory for quantum gravity, we need any model capable of formulating physical phenomena universally both in microcosm and macrocosm. As the first hypothesis, a topos of presheaves exists together with a t-site, which will be needed for Section III. 5. As we have noted, such presheaves become potentially observable only when they are reified with objects of the t-site, i.e., the presheaves must be in ur-particle states for being measured. The theory of t-topos should be developed in terms of the morphisms, e.g., factorability in the t-site and the associated t-topos to describe the natural phenomena of microcosm and macrocosm.

Section III. 2 Particle-Wave Duality and Ur-Entanglement

Our realization is that the presheaf representing a particle is a crucial entity whether the presheaf is reified or not. As we have already seen earlier, the particle-wave duality of a quantum entity is captured as the two faces of one object, i.e., the presenting presheaf for the particle. The consequences from our t-topos, as one expects, do not imply the deterministic views in microcosm even by considering the objects of the t-site as state determining variables in the classical sense of hidden variable theories. See, e.g., [18] as a survey article. The reason for the non-deterministic nature is in Axiom $(t-1)$. As for Axiom $(t-1)$, see the comments in [39] relating the work [10] and [11]. That is, a particle representing presheaf need not be defined for an arbitrary object of the t-site. In a naïve sense, the variables as t-site objects are hidden in the following sense. When a particle presented by a presheaf is observed (i.e., measured), then the associated presheaf is in an ur-particle state. However, the presheaf may be in an ur-particle state even when the presheaf is not observed. A much more detailed analysis of this matter will be studied in what will follow.

A measurement of a particle forces the presenting presheaf of the particle to be reified with an object of the t-site, which is a necessary condition for the presenting presheaf to be in an ur-particle state. However, as we have mentioned earlier, it is not a sufficient condition, i.e., the presheaf may be in an ur-particle state without being observed (measured). Such a classical quantum mechanical common phrase saying that a particle can *"be"* in several different locations at *"the same time"* needs to be reexamined whether such a phrase is appropriate and correct for describing the state of a quantum entity. Our formulation in terms of t-topos gives precise descriptions without such an ambiguity for such microcosmic issues. When taken literally in terms of the notions of the t-topos theory, such a loose expression as "An electron is everywhere" demands the existence of infinitely many reified objects of the t-site over which a representing presheaf must be defined. Such an ur-state is not allowed in t-topos. When one can use such expressions like *"be"* and *"the same time,"* we need to first know whether a presheaf associated with a particle is reified or not. Secondly, *"the same time"* means that in terms of a generalized time period, one object of the t-site has been chosen for several presheaves under consideration. At the level of a presheaf, there is no such thing as at *"the same time"*. Namely, time presheaf τ is locally defined for an object of the t-site. After all, the theory of temporal topos based on axioms $(t-0) \sim (t-4)$ (and the t-g, hypothesis, i.e., the t-topos theoretic gravitational hypothesis in Section III. 5 on black holes) is a study for qualitative quantum gravity. That is, for a particle to be at a place, first of all an object of t-site must be specified for the reification. Then correspondingly, our definition tells us that the particle must not be in an ur-wave state as long as an object in S has been specified. Namely, an expression as "An electron moves from point A to point B taking all available paths simultaneously" is assuming the following. First, the nearest possible place from point A can be defined by the corresponding ur-particle state to a micromorphism from the state for A. See III. 3 for the definition of a micromorphism. If such an electron were observed beside the two states corresponding to A and B, then there would be a non-trivial factorization of $V \xrightarrow{g} U$, e.g., $g = g_2 \circ g_1$ via $\{W\}$ in the t-side, corresponding to A and B. A non-trivial factorization of g means that neither g_1 nor g_2 is an isomorphism. Namely, in the diagram

$$\begin{array}{ccc} V & \xrightarrow{g} & U \\ & \searrow^{g_1} \nearrow^{g_2} & \\ & W & \end{array} \quad \text{(III. 2. 1)},$$

$V \xrightarrow{g_1} W$ and $W \xrightarrow{g_2} U$ would become proper (i.e., neither g_1 nor g_2 is an isomorphism) linearly t-ordered morphisms. In particular, if such a morphism $V \longrightarrow U$ is a micromorphism, by definition there does not exist such a proper factorization. Again, see III. 3 for a micromorphism. The t-topos formulation indicates that the number of such paths between A and B (linearly t-ordered) are precisely equal to the number of non-trivial factorizations by linearly t-ordered morphisms of $V \longrightarrow U$. If we assume that either g_1 or g_2 is not linearly t-ordered, then the ur-state at W would be outside the light cone with respect to V or U, respectively. See Section III. 5 for the t-topos theoretic notion of a light cone.

Next consider the t-topos theoretic formulation of the double slit experiment situation. See [41] for details concerning the t-topos version of the double-slit interference. Suppose that an electron gun is located equally distant from the two slits on a barrier. Electrons leave one by one with an appropriate interval keeping only one electron traveling from the electron gun through the barrier with two slits for a photographic plate as a detector screen. In the above, A corresponds to the location determined by object V of the t-site where an electron leaves the electron gun, and B corresponds to the spot determined by U on the photographic plate. As is the case of the usual setting of this experiment, there is a concentrated region on the photographic plate directly opposite the middle point of the two slits. We see the usual interference effects on the photographic plate away from the central region. The t-topos theoretic consequence is the following. When there is more than one slit (appropriately designed), such a t-site object W in the above diagram (III. 2. 1) cannot be uniquely determined. Therefore, such an electron presenting presheaf e is in an ur-wave state during the time periods corresponding to the two ur-particle states determined by V and U at the electron gun and the photographic plate, respectively. We examine this well-known experiment in terms of t-topos as follows. A typical argument about this quintessential experiment for the particle-wave duality is the following. First let O_1 and O_2 be the slits on the barrier. *Assumption 1: an electron goes through the barrier. Assumption 2: hence the electron* *goes through either O_1 or O_2.* We need to come back to these assumptions later for more than the obvious reasons. See the diagram below. Suppose that one electron went through O_1. Then usually one concludes that the electron did not go though O_2. This "either O_1 or O_2" logic does contradict such a diffraction pattern on the photographic plate. Then one usually concludes that each electron went through "both slits" based on Assumption 1. One does not usually say that each electron went through neither O_1 nor O_2 for the obvious reason, namely such an electron then never reaches the photographic plate of Assumption 1. The t-topos formulation becomes the following. The expression "went though either O_i, $i = 1, 2$" assumes the determination of a unique object of the t-site for either slit. Then it would be an ur-particle state for each slit. In other words, two objects from the t-site must be chosen for those two slits. Consequently, a unique object of the t-site cannot be chosen for the two slits O_i, $i = 1, 2$. In this case we have two choices for t-site objects for O_1 and O_2. Consequently, Assumption 2 is inappropriately phrased. The case of having more than one slit forces e not to be in an ur-particle state until the ur-particle state corresponding to U on the photographic plate is reached. Assumption 1, in terms of the t-topos formulation, can be phrased as follows. When there is a barrier with more than one slit, then there is more than one choice in the t-site objects to be ur-particle states. Assumption 2 would become: two slits make it impossible to specify a unique object of the t-site for the barrier with more than one slit. In other words, the electron approaches the photographic plate being in an ur-wave state because of the non-uniqueness in choice of an object of the t-site for the barrier. We observe from our point of view that since there does not exist a specified object when "passing through" a slit, there is not a linearly t-ordered relation from either of the history to the time reified at the photographic plate.

The delayed choice experiments can be understood in a similar manner. Suppose that a single photon represented by presheaf γ splits at G into two paths which are reflected by the mirrors O_1 and O_2 meeting at D. Let S be a switch between O_2 and D. Switch S is letting a photon go through or turning aside to a detector P. Let V be the object of the t-site corresponding to the ur-particle at G. After leaving G, these two paths correspond to the factorization

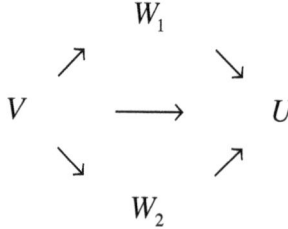

of the morphism $V \to U$ where W_1 and W_2 correspond to paths O_1 and O_2, respectively, and U corresponds to D. Hence, since a unique choice of an object in the t-site cannot be made, γ is in an ur-wave state after leaving G. When S lets a photon go though, γ is in an ur-wave state until the ur-particle state is established corresponding to U (because of the non-uniqueness of factorization). When the detector P does not resister the photon, the factorization has no choice but $V \to W_1 \to U$. Namely, γ is in an ur-particle state. It is irrelevant where the detector is set in this apparatus.

However, whether there is more than one choice from the t-site or not is relevant to determining a presheaf being either in an ur-particle state or in an ur-wave state in the t-topos sense. In other words, in t-topos theoretically speaking, a particle-associated presheaf reified at the photographic plate does not have two different histories after leaving the electron gun. The presheaf was simply in the ur-wave state between the photographic plate and the electron gun. It is the wrong usage of a word in the t-topos sense to say an electron "went (goes) through the barrier with the two slits." This is because to be able use the word "going through a slit" an object of the t-site needs to be chosen. The well-known Feynman's dendenkenexperiment for determining whether an electron going through one of the slits or not is simply the question of the uniqueness of the choice between the two t-site objects W_1 and W_2.

One of the Einstein-Podolsky-Rosen (referred to as EPR) experiment interpretations is the following. Let γ_1 and γ_2 be the presheaves representing two photons created by a decay of a certain atom moving in opposite directions from the decay spot. If γ_1 is measured by P over a generalized time period V, we have a morphism $\gamma_1(V) \xrightarrow{S_V} P(V)$. One can treat the relativistic case. However, for the sake of simplicity, we will first give a formulation to the non-relativistic case. See Section III. 5 for the relativistic treatment. The notion of an ur-entanglement is independent of the distance between γ_1 and γ_2, hence we can formulate when γ_1 and γ_2 are in *mutually non-intersecting universal light cones* (See Definition III. 5. 4.), which may be regarded as our non-locality property of an ur-entangled pair. We need a proper definition so that there exists a correlation between the ur-state of γ_1 which P observes and the ur-state of γ_2. For the observation morphism $\gamma_1(V) \xrightarrow{S_V} P(V)$, there does not exist a morphism $\gamma_2(V) \xrightarrow{S_V} P(V)$. This is because one cannot compose the morphisms $\gamma_1(V) \xrightarrow{S_V} P(V)$ and $\gamma_2(V) \longrightarrow \gamma_1(V)$ for mutually non-intersecting universal light cones. See preceding Remark III.1.2. (4). However, when γ_1 and γ_2 are entangled, the ur-particle state of γ_2 is determined as t-site simultaneously as γ_1 is measured, where t-site simultaneity means with the same ur-state determining t-site object. See the following diagram:

$$\gamma_1(V) \leftarrow ----- \gamma_2(V)$$
$$\downarrow$$
$$P(V)$$

where the morphism $\leftarrow\text{-----}$ indicates the non-existing morphism outside the mutually intersecting light cones. This observation leads to the following t-topos theoretic definition of (ur-) entanglement.

Definition III. 2. 1 Let p and q be presheaves representing two particles. Then p and q are said to be *ur-entangled* if a pair (p,q) of presheaves acts as one presheaf in $\hat{S}$. Namely, the definition of an ur-entangled pair of presheaves p and q in t-topos is: presheaf p is defined for a t-site object V if and only if q is defined for V. Then we define the reified state of the pair as $(p,q)(V) = (p(V),q(V))$. One can say that an ur-entangled reified pair of p and q is in a *t – site simultaneous ur-particle state*. The concept of t-site simultaneity is the only simultaneity there can be in any sense.

An obvious consequence is that for ur-entangled presheaves p and q, p is in an ur-wave (or ur-particle) state if and only if q is in an ur-wave (or ur-particle) state. Suppose two entangled particles Q and Q' represented by presheaves p and q are a distance apart from each other. If presheaf q is observed by P over a generalized time period V, we have a morphism $q(V) \xrightarrow{S_V} P(V)$. For example, consider a pair of spin $\frac{1}{2}$ – particles Q and Q' represented by ur-entangled presheaves p and q of a combined spin zero state. If the measured spin of Q is *up* in a certain direction, then the spin of Q' is *down* in that direction. Then the other presheaf p of the ur-entangled pair becomes *t-site simultaneously* in an ur-particle state evaluated at V by Definition III. 2.1. As noted earlier, there does not exit a canonically induced morphism to $P(V)$ from $p(V)$.

Remark III. 2. 2 Let p and q be ur-entangled. Then Definition III. 2. 1 implies that one cannot determine whether p and q are ur-entangled or not unless either of p or q (hence both p and q) is in an ur-particle state. Another consequence from our formulation is that a measurement of p over an object V of the t-site determines the ur-particle state $q(V)$ of q.

Epitome III. 2 Let us consider an electron which is not measured (observed). The classical quantum interpretation of such an electron is that the electron is said to spread out, i.e., in the superposition. This is the case in terms of t-topos when the presheaf e associated with the electron is not evaluated at any object of the t-site S. Namely, presheaf e is not reified, i.e., e is in an ur-wave state (or in an ur-superposition state). See Definition III. 1.1 $(t-1)$. According to the t-topos formulation, it is inappropriate to use such a phrase as: an electron is in the superposition of being everywhere. In the sense of t-topos, such an electron is "nowhere" may be the better approximation to describe such a non-reified presheaf e. This is because "being here" implies the existence of a t-site object with which e is reified. Here is another way the t-topos theoretic interpretation of an electron is in such a situation: even though there exists a unique morphism from e to space-time sheaf $\omega = (\kappa, \tau)$, however, without an object of the t-site the electron associated presheaf e and $\omega = (\kappa, \tau)$ are both in an ur-wave state. In particular, one has no information about the location of the electron in terms of $\omega = (\kappa, \tau)$. This kind of commonly used phrase is an opposing categorical description to the t-topos theoretic formulation of superposition of being a non-reified presheaf. In the case of the double slit experiment discussed earlier, let $V \to W_1 \to U$ and $V \to W_2 \to U$ be the possible factorizations of $V \to U$ corresponding to the two possible ur-particle states of e at the barrier with two slits between firing an electron and reaching the photographic plate. When $V \to W_1 \to U$ is chosen by observing the electron going through slit O_1, the probability of going through O_2 becomes zero. Note that there does not exist a linearly t-ordered morphism between W_1 and W_2. This jumping of the quantum ur-state is corresponding to the collapse of the wave function to one of the eigenstates. After passing though O_1, e returns to an ur-wave state, i.e., continuous Schrodinger evolution, until e assumes the ur-particle state corresponding to U on the photographic plate.

Another related thought experience example is about a box containing one electron, whose associated presheaf is denoted again by e. The box is divided into two parts. Our t-topos interpretation is as follows. With the very action of separating the box, e.g., by cutting the box, there is a moment to consider the electron going through one slit rather than multiple slits. That moment forces the presheaf e to be in an ur-particle state, thereby determining in which part of the box the electron is located. Hence, for example, there is no need to separate the two parts of the box in a long distance, i.e., such an action is irrelevant. Namely, this thought experiment is a distinct concern from the quantum entanglement discussed.

Section III. 3 Micromorphism; Uncertainty Principle and Quantum Tunneling

The notion of a micromorphism is one of the most crucial notions in our theory of temporal topos approach to quantum gravity. We will give a definition of a micromorphism in what will follow as a non-factorable morphism in the t-site. In the case of a linearly t-ordered case, a most immediate state of a given state is provided by a linearly t-ordered micromorphism from the given state. One of the most crucial main pillars of quantum physics and also for our t-topos is the Heisenberg uncertainty principle in observations. Namely, two observables cannot be simultaneously observed with exact accuracy. Such a notion as a micromorphism in terms of t-topos notions is to give a more fundamental reason behind the well known version of the Heisenberg uncertainty principle. We will give the t-topos theoretic descriptions why and when quantum tunneling can occur. Recall that classically the uncertainty principle of Heisenberg is expressed as the product $\Delta x \cdot \Delta p$ (or $\Delta E \cdot \Delta t$) never less than the Planck constant $\hbar$, where Δx (or ΔE) is the uncertainty in position (or energy) and Δp (or Δt) is the uncertainty in momentum (or time) which is the consequence of two Hermitian operators satisfying the canonical commutation relations for the self-adjoint operators for coordinates and momenta. That is, the Heisenberg uncertainty principle is derived from a Schrodinger equation showing the wave property of a particle. Namely, the uncertainty comes from the wave property, i.e., two observables, the Hermitian operators, can simultaneously be measured with only restricted accuracy. The following is the fundamental and crucial definition in the t-topos theoretic methods. We will give a description of the Heisenberg uncertainty principle based on the notion of a micromorphism, which is relevant to the question such as when and why a quantum tunneling can (and must occur). With the notion of a microdecomposition and Axiom $(t-4)$ in the next section III.4, quantum fluctuations can be formulated in terms of t-topos.

Definition III. 3.1 A morphism of objects $\zeta : X \longrightarrow Y$ in the t-site S is said to be a *micromorphism* if $\zeta : X \longrightarrow Y$ cannot be factored by proper morphisms in the following sense. If $\zeta = \alpha \circ \beta$ holds, where $\beta : X \longrightarrow W$ and $\alpha : W \longrightarrow Y$ are morphisms in S, then either α or β is an isomorphism of S. In other words, when there exists a commutative diagram as

$$\begin{array}{ccc} X & \xrightarrow{\zeta} & Y \\ & \searrow^{\beta} \quad \nearrow_{\alpha} & \\ & W & \end{array} \qquad \text{(III. 2. 2)}$$

W is isomorphic to either X or Y in the t-site S. Note also that the notion of a linearly t-ordered micromorphism can be defined. Namely, a linearly t-ordered morphism $X \xrightarrow{\zeta} Y$ is said to be a *linearly t-ordered micromorphism* when for a factorization by linearly t-ordered morphisms as in the above, either α or β is an isomorphism.

Definition III. 3. 2 Let $f : V \longrightarrow U$ be an arbitrary morphism of objects in S. A sequence $V = V_0 \xrightarrow{\zeta_0} V_1 \xrightarrow{\zeta_1} V_2 \xrightarrow{\zeta_2} \mathrel{-}\mathrel{-}\mathrel{-}\mathrel{-} \xrightarrow{\zeta_{n-1}} V_n = U$ of morphisms and objects is said to be a *microfactorization* of $f : V \longrightarrow U$, if for the factorization $f = \zeta_{n-1} \circ \mathrel{-}\mathrel{-}\mathrel{-}\mathrel{-} \circ \zeta_2 \circ \zeta_1 \circ \zeta_0$, each ζ_k is a micromorphism, $0 \le k \le n-1$. Similarly, for a linearly t-ordered morphism, one can also formulate the notion of *linearly t-ordered pure microfactorization* by insisting that every ζ_k is linearly t-ordered. Intuitively speaking, the larger the number n of a linearly t-ordered factorization of $V \xrightarrow{f} U$ becomes, the longer the time period between the two ur-states corresponding V and U is. For a specified particle representing presheaf m, the factorization number $n = n_m$ can be affected if m is representing a non-zero mass particle because of the t-g. hypothesis in Section III. 5. One can say that a morphism $V \xrightarrow{f} U$ is an n –*micro generalized time period morphism* for m. Then the non-negative integer $n = n_m$ is said to be the t-*microlength* of $V \xrightarrow{f} U$ of m. For the photon representing presheaf case $m = \gamma$, see Notes III. 6. 9 (1).

Note III. 3. 3 (1) In the case of measurements, the notion of a linearly t-ordered microfactorization is crucial. Namely, temporally closest measurable ur-particle states are provided by a linearly t-ordered micromorphism between the ur-particle states. In other words, if $m(V) \to m(U)$ are the corresponding ur-particle states to a linearly t-ordered micromorphism $V \to U$ for a presheaf m, then it is impossible for m to be in an ur-particle state between the two states $m(V)$ and $m(U)$ by the definition. That is, m has to be in an ur-wave state between two ur-particle states $m(V)$ and $m(U)$. Note that as we shall see in what will follow, during the time interval between $\tau_m(V)$ and $\tau_m(U)$ corresponding to the linearly t-ordered micromorphism $V \to U$, quantum tunneling for m can occur. In general, the notion of a micromorphism, which need not be linearly t-ordered, is often used for the analysis of a presheaf representing a macro entity, e.g., for a microdecomposition (together with a microcovering of a t-site object) of the presheaf in the following section.

(2) For a presheaf m and for a micromorphism $V \xrightarrow{g} U$ in the t-site S, let the corresponding ur-particle states be defined $m(V)$ and $m(U)$. Then we have the induced morphism $m(U) \xrightarrow{m_g} m(V)$, for the corresponding ur-particle states. We also assume that $V \xrightarrow{g} U$ is linearly t-ordered. Then by the definition of a micromorphism, as we saw in the above, m must be in an ur-wave state between the ur-particle states $m(V)$ and $m(U)$. This implies that between the time intervals $\tau_m(V)$ and $\tau_m(U)$, it is impossible to localize the position (and time) of m because of the very definition of a micromorphism. Explicitly described: suppose that we could find the position of m before $\tau_m(U)$ and after $\tau_m(V)$. Let W be the corresponding object in the t-site S corresponding to the ur-particle state of m determined by the position. Namely, we have $V \xrightarrow{g_1} W$ and $W \xrightarrow{g_2} U$. Then $V \xrightarrow{g_1} W \xrightarrow{g_2} U$ becomes a factorization of the linearly t-ordered morphism $V \xrightarrow{g} U$. This contradicts the assumption of the morphism g being a micromorphism. Consequently, it is inappropriate to use a commonly used phrase like "the particle *is everywhere* between the two ur-particle states" that are determined by the generalized time periods V and U. This is because in order to be able to use the word "be" anywhere, an object in the t-site S needs to be specified. But such a specification of an object of the t-site could not be the case for a micromorphism. Note also that this observation of a micromorphism plays a crucial role in quantum tunneling in what will follow. In terms of space-time sheaf ω, we can formulate as follows. For the unique morphism $m \xrightarrow{\sigma_m} \omega$, such a micromorphism $V \to U$ as above induces the diagram

$$\begin{array}{ccc} m(V) & \xrightarrow{\sigma_m(V)} & \omega(V) \\ \uparrow & & \uparrow \\ m(U) & \xrightarrow{\sigma_m(U)} & \omega(U) \end{array} \qquad (\text{III.2.3})$$

Then the vertical morphisms are canonical morphisms induced from the micromorphism $V \to U$ by the presheaves m and ω. In particular, the morphism $\omega(U) \xrightarrow{\omega(g)} \omega(V)$ also cannot be factored. Hence one cannot measure the space-time ur-state for the presheaf representing a particle during the time period corresponding to the micromorphism $V \to U$.

(3) The classical usual time interval induced between $\tau(V)$ and $\tau(U)$ for a morphism $V \to U$ corresponds to Planck time if $V \to U$ is a linearly t-ordered micromorphism. Namely, this is the shortest time period one can assign a scalar for two different linearly t-ordered ur-particle states corresponding to generalized time periods V and U. Such a definition of a Planck time should be given as in the above so that all the relevant notions may become consistent within the t-topos theory. As noted earlier, besides the fields $\mathbb{R}$ of real numbers and $\mathbb{C}$ of complex numbers, an assignment of a scalar of a physical microcosmic entity measurement such a field as p-adic numbers has been considered. See for example [78] for $\hat{\mathbb{Q}}_p$ in his p-adic string theory. We highly recommend [47] as a reference for p-adic analysis and related topics which may be more approachable than [9]. Notice that the above definition of a shortest measurable period is defined without a specified presheaf representing a particle. However, such a shortest time period between $\tau(V)$ and $\tau(U)$ may not be universal in the following sense. A question one must ask is whether such a notion as Planck time is a presheaf-independent or not. If it is universal, i.e., independent of a presheaf, then such a shortest time period should be called *the* Planck time rather than *a* Planck time. For such a universality a photon presheaf can be used (See Section III. 5.). Let $V' \to U'$ be another linearly t-ordered micromorphism. To be explicit, the time interval between $\tau(V')$ and $\tau(U')$ may be different from the one for $\tau(V)$ and $\tau(U)$ where $U \neq U'$. The dependency or independency of the notion of a micromorphism on a presheaf has to be handled with care. Let $V \to U$ be a linearly t-ordered micromorphism where m is reified as an ur-particle state $m(V)$. Then the difference of the ur-states $\omega(U)$ and $\omega(V)$ of the associated space-time may not be universal, i.e., dependent of a presheaf m effect on space-time $m(V) \to \omega(V)$. This concern about the universality still needs to be examined in future study; namely whether one can consider the notion of a micromorphism as a universal unit by assigning a unit for 1 t-microlength in Definition III. 3. 2, or not.

The Heisenberg uncertainty principle is one of the main pillars for characterizing the nature of quantum microcosm. Commonly, it is phrased as the impossibility of measuring simultaneously both the position and the velocity (or momentum) of a particle with exact accuracy. We will formulate a t-topos theoretic

Heisenberg uncertainly principle as an uncertainty to measure the state and its change in the t-topos sense. We hope that our approach provides a more accurate structural characterization and reveals a truer nature of this fundamental uncertainty principle. We will also give the definition of a fundamental presheaf in terms of a direct limit in Definition III. 4.1. A fundamental presheaf is a presheaf which corresponds to an elementary particle. In our terminology, the particle presented by a fundamental presheaf is said to be t-elementary. (See Remarks III. 4.5.)

III. 3. 4 Uncertainty Principle in t-topos

As we will define the notion of a microdecomposition of a presheaf in Section III.4, a presheaf m can be decomposed into subsheaves $\{m_j\}_{j \in J}$ so that $m = \prod_{j \in J} m_j$ becomes a microdecomposition. This means that after a microdecomposition of a presheaf m, each component m_j represents a microcosmic particle. Let a presheaf m represent an elementary particle. We will discuss the principle of uncertainty for a microcosmic object. Namely, we consider a fundamental presheaf m. (See Definition 4.1. for the definition of a fundamental presheaf.) Let $V \longrightarrow U$ be a linearly t-ordered micromorphism corresponding to ur-particle states $m(V)$ and $m(U)$. Suppose that those two ur-particle states corresponding to generalized time periods V and U are observed. Then for the linearly t-ordered micromorphism, let the corresponding ur-particle states of space-time sheaf ω_m associated with the presheaf m be $\omega_m(V) = (\kappa_m, \tau_m)(V) = (\kappa_m(V), \tau_m(V))$, and $\omega_m(U) = (\kappa_m, \tau_m)(U) = (\kappa_m(U), \tau_m(U))$. Then the t-topos theoretic uncertainty principle involves the following two kinds of uncertainty.

(U. P. 1) For the linearly t-ordered micromorphism $V \longrightarrow U$, the states $\omega_m(V)$ and $\omega_m(U)$ of space-time induced by space-time sheaf ω corresponding to V and U are the nearest ur-particle states of m with respect to space-time sheaf ω. That is, between the classical linearly ordered times $\tau_m(V)$ and $\tau_m(U)$ of m, presheaf m must be in an ur-wave state in the sense that the particle represented by presheaf m must be in a superposition. See the Heisenberg commutative diagram in the following.

The *maximum* of a distance determined the two ur-particle states corresponding to V and U occurs when m is a photon presheaf γ. That is, the measured value for the difference of two states, evaluated in an appropriate scalar field K of the states $\omega_\gamma(V)$ and $\omega_\gamma(U)$ during the intervals $\tau_\gamma(V)$ and $\tau_\gamma(U)$, corresponds to such a maximum. Therefore, there exists a number $\varepsilon > 0$ (referred to as the Planck length) in an appropriate field, e.g., in $K = \mathbb{R}$ or in $K = \hat{\mathbb{Q}}_p$ such that the difference between the physical quantities, e.g., the uncertainty in position corresponds to the distance $c \cdot |\tau(U) - \tau(V)| \geq \varepsilon$, where c is the speed of light, and $\tau(U) - \tau(V)$ indicates the difference of the ur-particle states of τ. See Isham-

Butterfield in [10], [11] for assigning such a number $|\tau(U)-\tau(V)|$ by the Fundamental Composition Principle (referred to as FUNC in that paper). As the scalar field, the field $\mathbb{R}$ of real numbers has been employed. As a (semi-) norm having values in $\mathbb{R}$, one could use, e.g., a non-Archimedean norm (e.g., p-adic norm), where the value of an element $x + y$ of any normed ring does not change from the value of x until the norm of y becomes larger than that of x. See [9] for non-archimedean analysis and see [78] in the p-adic approach for quantum gravity.

Note that the difference of two states $\kappa(U)$ and $\kappa(V)$ corresponds to Δx in the classical formulation of Heisenberg uncertainty principle $\Delta x \cdot \Delta p \geq h/4\pi$ where h is Planck constant. That is, $h = 2\pi\hbar$ which is about $2\pi 10^{-34} J \cdot s$.

(U. P. 2) The second uncertainty comes from the very nature of our categorical approach. That is, the Dedekind-Cantor continuum (as the measured value field) approaches to a physical entity should be avoided for a proper treatment of a singularity. A point versus an object (with thickness) view corresponds to a number versus an interval, when $\mathbb{R}$ of real numbers is used as the value field. The position and time corresponding to κ and τ take the interval-values for objects of the t-site whose assigned sizes are depending upon the objects in t-site S and the presheaves. Namely, corresponding to two objects in the t-site, one can only assign two intervals, e.g., contained in the field $\mathbb{R}$, rather than two real numbers, to the quantities for the objects $\omega(V)$ and $\omega(U)$. This aspect to uncertainty is irrelevant to the matter of the measured difference of the corresponding ur-particle states induced by a micromorphism, but any ur-particle state of a presheaf must have the value for the "thickness" of the state rather than a point like treatment.

The most direct t-topos theoretic presentation of uncertainty principle (abbreviated as TTUP) can be phrased in terms of the ur-state and the micromorphism.

(TTUP) To determine an ur-particle state of m is to specify an object V of the t-site. The uncertainty occurs in the sense of the above (U.P.2) for an assignment by a measurement in one of the categories in $\prod_{\beta \in \Lambda} C_\beta$ for the scale of $m(V)$ taking an interval, e.g., in $\mathbb{R}$ or $\hat{\mathbb{Q}}_p$. Then, the next nearest possibly measurable ur-particle state of m is determined by a linearly t-ordered micromorphism $V \longrightarrow U$ and by an object U of the t-site. The change of these ur-particle states corresponding to the linearly t-ordered micromorphism $V \longrightarrow U$ is the uncertainty between the two ur-particle states $m(V)$ and $m(U)$. That is, the uncertainty for the ur-state of m in the t-topos sense is defined as the gap of those two ur-particle states corresponding to a micromorphism without referring to ω as in (U. P. 1) and (U. P. 2).

For example, in order to determine the rate of any change in ur-particle state of m, one needs at least two ur-particle states corresponding to such a linearly t-ordered micromorphism. Between the time interval from $\tau_m(V)$ and $\tau_m(U)$ it is impossible to specify the position of m since m is not allowed to be in an ur-particle state, i.e., no proper factorizations of $V \longrightarrow U$ can exist for a possible ur-particle state.

III. 3. 5 The Heisenberg commutative diagram

Let $V \xrightarrow{g} U$ be a linearly t-ordered micromorphism where a presheaf m is reified at V and U. Furthermore, suppose that m is measured by P over V and U. The following commutative diagram

(H.C.D)

$$\begin{array}{c} m(V) \xleftarrow{m(g)} m(U) \\ \downarrow \quad\quad \downarrow \\ \omega(V) \xleftarrow{\omega(g)} \omega(U) \\ \downarrow \quad\quad \downarrow \\ P(V) \xleftarrow{P(g)} P(U) \end{array}$$

summarizes t-topos formulation for the uncertainly principle. Note that $V \xrightarrow{g} U$ is just one of several possibly reifiable linearly t-ordered micromorphisms with the domain object V. Note also that it may be a worthwhile investigation to unify the above Heisenberg commutative diagram with relativistic diagrams (III. 5. 5.) and (III. 5.6) in Section III. 5. In the Heisenberg commutative diagram (H. C. D.), for example, the morphism $m(V) \xleftarrow{m(g)} m(U)$ should be read as follows: as indicated earlier, the morphism $m(g)$ indicates the change of ur-states of m during the generalized time periods determined by the micromorphism $V \xrightarrow{g} U$. The commutativity of the portion

$$\begin{array}{c} m(V) \xleftarrow{m(g)} m(U) \\ \downarrow \quad\quad \downarrow \\ \omega(V) \xleftarrow{\omega(g)} \omega(U) \end{array}$$

measures, via $\omega(V) \xleftarrow{\omega(g)} \omega(U)$, i.e., by the change of ur-states of the space-time sheaf, the effect on the space-time sheaf ω by the mass of m in the classical form of the Heisenberg uncertainty principle. See the t-g. hypothesis in Section III. 5.

Remark III. 3. 6 (1) In the case of a macro object presheaf m, let $m(W)$ and $m('W)$ be the ur-particle states of m corresponding to a linearly t-ordered morphism $W \xrightarrow{g} 'W$. Heisenberg's uncertainty principle still applies to the macro case. However, as we have already discussed, there are (too) many possible factorings of the linearly t-ordered morphism $W \xrightarrow{g} 'W$ into micromorphisms, and also possible microdecompositions of m (See III. 4.). Namely, for a macro object (presheaf), the effects of uncertainty principles are *too delicate* to be observed with respect to macro level morphisms and objects.

(2) The flow of information from microcosm to macrocosm is functorial as we will explain in Remarks III. 4. 5. (2). However, from macrocosm to microcosm the situation is highly non-functorial (non-canonical). Namely, the induced morphism $m(W) \xleftarrow{m(g)} m(W')$ by $W \xrightarrow{g} {'W}$ in macrocosm is the totality induced from the factorizations $\{{_0}W_i^j \to {_1}W_i^j \to ---\to {_k}'W_{i'}^j\}_{j \in J}$ of ${_0}W_i^j \longrightarrow {_k}'W_{i'}^j$ associated with all m_j, $j \in J$, where $m = \prod_{j \in J} m_j$ is a decomposition of m, and where the linearly t-ordered morphism $W \xrightarrow{g} {'W}$ induces a covering $\{W^j \longleftarrow {_0}W_i^j\}_{i \in I_{W^j}}$ of W^j and a covering $\{'W^j \longleftarrow {_k}'W_{i'}^j\}_{i' \in I_{'W^j}}$ of $'W^j$ with which m_j may be reified.

This observation of the macro level Heisenberg's uncertainty principle is relevant to sections III. 5 and III. 6 in what will follow.

(3) We will introduce the notion of a fundamental presheaf which is basically a presheaf presenting an elementary particle. (See the paragraph preceding III. 3. 3 and see Definition III. 4. 1 for the definitions of a fundamental presheaf and a fundamental t-site object.) Next we will consider the following objects $\{V\}$ of the t-site. Suppose that a fundamental presheaf m is reified at a fundamental object V of the t-site S. However, there does not exist any U so that $V \longrightarrow U$ becomes a linearly t-ordered morphism and where m is reified as an ur-particle state $m(U)$. Then, one may interpret those time periods $\tau_m(V)$ for such t-site objects as borrowing energy from coming and going vacuum excitation. (See the comment in the following Quantum Tunneling for the case where only a micromorphism $V \to U$ from V to U exists.) Namely, the states where only fundamental presheaves with fundamental objects without having any t-linearly ordered morphisms are the lowest energy states, i.e., the vacuum energy states. Namely, I will give the following definition of a virtual state of a particle in the sense of t-topos. Let m be a fundamental presheaf. Then m is said to be in a *t-virtual particle state* at a fundamental object V when

(i) Presheaf m is reified at V,

and

(ii) there does not exist a t-linearly ordered morphism from V to any objects of the t-site S.

Then such an object V of the t-site from which any t-linear morphism is not defined to another object as above is said to be a *t-virtual object* (a *virtual object*, or a *t-virtual generalized time period*) of the t-site S. Compare with the definition of the ur-bigbang of the $(-1)^{st}$ stage in Section III. 5. Note that at a virtual object V, the universal light cone for m collapses to a single object $m(V)$. (See Definition III. 5.2

for the notion of a universal light cone.) Because of the t-g. hypothesis in Section III. 5, this description of the above t-virtual particle state is relevant to the description of a black hole of mass (the Planck mass) $2 \times 10^{-5} g \sim \sqrt{\frac{\hbar c}{G}}$ caused by the energy uncertainty coming from classical Heisenberg's uncertainty principle, (here ignoring 4π), when observed in a micro scale of the Planck length $1.6 \times 10^{-33} cm \sim \sqrt{\frac{G\hbar}{c^3}}$, i.e., Schwarzchild's radius. With the last axiom $(t-4)$ of t-topos in the next Section III. 4, one may be able to imagine a topologically varying quantum foam-like state in the Planck scale if one views such a microcosm from the notions of a microdecomposition and a microcovering in the next section.

Note III. 3. 7 The reader may be able to have certain t-topos theoretic views by now from the established concepts thus far about the microcosm realm. Based on previously defined notions, we will formulate possible various stages relevant to the classical notion of a big bang in III. 5 from which linearly t-ordered morphisms have reached the present. The notion of a t-virtual particle state defined above is also relevant to the notions of the ur-big bang of the various stages where there are no t-linearly ordered morphisms leading to the present time. See what will follow.

III. 3. 8 Quantum Tunneling

The interplay among Heisenberg's uncertainty principle (the limits to the complementary nature for measurements, i.e., the above U.P.1 and U.P.2), superposition (the ambiguity in position and velocity of an un-measured entity during the interval between the two ur-particle states corresponding to a (micro) morphism), and the concept of a micromorphism (i.e., closest possible states in terms of a morphism of the t-site), gives a description of a quantum tunneling. Generally speaking, for a presheaf m representing a particle for which $m(V)$ is defined, and for a linearly-ordered morphism $V \xrightarrow{g} U$, the closer g is to a micromorphism, the more likely a quantum tunneling can occur between the two ur-particle states corresponding to V and U. However, whether a quantum tunneling must occur or not is a t-site object sensitive matter. Namely, under the same condition, for a given object V of the t-site, there may be more than one micromorphism having V as the domain t-site object. When there exists only one micromorphism from V, a quantum tunneling must occur in the sense that m cannot be in an ur-particle state until the ur-particle state $m(U)$ is reified at the codomain t-site object U for a micromorphism $V \to U$, but if there is more than one micromorphism from V, a quantum tunneling may not occur for a given barrier.

According to the t-topos theoretic interpretation of the uncertainty principle for a linearly t-ordered micromorphism $V \to U$, one can say that some extra energy may be borrowed within the difference of $\tau_m(U)$ and $\tau_m(V)$. In principle, according to our approach, it is not because of insufficient kinetic energy for a phenomenon of quantum tunneling not to occur, but because it is the microcosmic nature of the linearly t-ordered morphism. Consider the case when $m(V)$ is the ur-particle state before the transmitted ur-state over a potential barrier, and $m(U)$ is the transmitted ur-particle state. Suppose that $V \xrightarrow{g} U$ is a linearly t-ordered micromorphism. Then by definition, the morphism g cannot be factored by linearly t-ordered morphisms. The time gap between $\tau_m(U)$ and $\tau_m(V)$ is the uncertainty in time for the linearly t-ordered micromorphism $V \xrightarrow{g} U$. That is, as mentioned before, m is not allowed to be in an ur-particle state between the ur-states $m(V)$ and $m(U)$ corresponding to the generalized time periods V and U, respectively. During the time periods between $\tau_m(V)$ and $\tau_m(U)$, one commonly phrases a tunneling as: borrowing needed energy, i.e., the uncertainty against time beginning from the time $\tau_m(V)$ with the prompt payback by the time $\tau_m(U)$ after tunneling the barrier.

Epitome III. 3 The t-topos interpretations of Heisenberg's uncertainty principle and the duality in quantum physics are coming from:

(i) the notion of a micromorphism

and

(ii) whether a presheaf is reified with an object of the t-site or not.

Consequently, for the two generalized time periods corresponding to the domain V and the codomain U of a micromorphism $V \xrightarrow{g} U$, the t-topos theoretic formulation implies that no position and time can be specified, i.e., in terms of the space-time presheaf ω. Namely, the unique morphism $m \to \omega$ cannot be evaluated at any object of the t-site during the ur-wave state between the generalized time periods V and U. The uncertainty principle of Heisenberg can be viewed more explicitly as follows. Let m be a presheaf associated with a micro-object particle (e.g., an electron, or more generally, any t-elementary particle represented by a fundamental presheaf, which will be defined in the next section). For a micromorphism $V \to U$, let $m(V) \xrightarrow{\sigma_m(V)} \omega(V) \xrightarrow{s_V} P(V)$ be a measurement morphism over V of m by P via space-time presheaf ω and similarly for U. One can phrase the t-topos version of the uncertainty principle (TTUP) as the difference of the composition morphisms $s_U \circ \sigma_m(U)$ and $s_V \circ \sigma_m(V)$ over the time periods determined by the micromorphism $V \to U$. The ambiguity rooted on a micromorphism $V \to U$ and the associated measurements via space-time presheaf ω combines the both uncertainty formulations with respect to the measurements of energy and time and moment and position. Note that the unique morphism σ_m from m space-time ω provides the measurements of, e.g., the mass of the presheaf m over the micromorphism $V \to U$. Recall also that the t-topos uncertainty principle has another aspect, namely, (U.P.2) in the above, i.e., coming from the non-point (non-Dedekind-Cantor type) categorical formulation itself.

For an ur-particle state $m(V)$, as was mentioned briefly earlier, the non-uniqueness of linearly t-ordered morphisms from V to another objects $\{U\}$ of the t-site is the probabilistic aspect of the quantum physics. However, for a non-linearly t-ordered morphism $V \xrightarrow{g'} V'$, the ur-particle state $m(V')$ would be outside the light cone with respect to $m(V)$. See Definitions III.1 and III.2 for the t-topos version of a light cone.

Section III. 4 Microdecompositions and Microcoverings; Fundamental Presheaves Particles

One of the main issues in quantum mechanics is the connecting mechanism between microcosm and macrocosm. See Remark III. 3. 5. We will give a formulation for how the macrocosm measurement and the microcosm measurement are related in terms of decompositions of a presheaf and coverings of a generalized time period. We will express Planck scales in terms of the notions of inverse (projective) and direct (inductive) limits for our study. For an inverse system and a direct system, see Definition I. 1.6. A certain aspect of the interplay between quantum entities and global macro entities is sometimes referred to as a decoherence issue. We are going to decompose a presheaf m into finer presheaves as $m = \prod_{j \in J} m_j$. Furthermore, we will consider a decomposition of each component m_j of m to obtain even finer presheaves $m_j = \prod_{j \in J} m_{j_k}$. Then we get a sequence of decompositions of presheaves as follows:

$$m = \prod_{j \in J} m_j \longrightarrow m_j = \prod_{j \in J} m_{j_k} \rightarrow ---, \qquad \text{(III. 4. 1)}$$

where each morphism of the direct system in (III. 4.1) is the projection. Note that this is the sequence that appeared in (3) of Notes I. 3. 5.

Next, for an object V of a t-site S, consider a covering $\{V \longleftarrow V_i\}_{i \in I}$ of V. Furthermore, consider a covering $\{V_i \longleftarrow V_{i_l}\}_{l \in L}$ of V_i, where all the index sets are finite sets. Then we have the following inverse system of the coverings:

$$\{V \longleftarrow V_i\}_{i \in I} \longleftarrow \{V_i \longleftarrow V_{i_l}\}_{l \in L} \longleftarrow ---, \qquad \text{(III. 4. 2)}$$

where each morphism is induced by the compositions of covering morphisms. For this observation, see (G.3) of Definition I. 3.1. Or, we get also

$$\{V \longleftarrow V_i\}_{i \in I} \longrightarrow \{V_i \longleftarrow V_{i_l}\}_{l \in L} \rightarrow --- \qquad \text{(III. 4.3)}$$

where each morphism is also induced by the compositions of covering morphisms. See Definition I. 3.1 for coverings.

We are now ready to describe the postponed last axiom of t-topos:

(t-4) For each covering of an object of the t-site $\{V \longleftarrow V_i\}_{i \in I}$, the corresponding time period assigned to $\tau(V)$ is not shorter than that of $\tau(V_i)$.

Definition III. 4. 1 A presheaf m is said to be a *fundamental presheaf* when every decomposition of m consists of only isomorphic presheaves to m. Namely, for a fundamental presheaf m, an arbitrary decomposition of $m = \prod_{j \in J} m_j$ implies that each m_j is isomorphic to m as presheaves. For a decomposition $m = \prod_{j \in J} m_j$, if each presheaf m_j is a fundamental presheaf, such a decomposition is said to be a microdecomposition of m.

Definition III. 4. 2 An object V of the t-site S is said to be a *fundamental object of the t-site S* when any covering $\{V \leftarrow V_i\}$ of V consists of objects isomorphic to V. Such a covering is said to be a *microcovering*.

Definition III. 4. 3 The direct limit of the sequence (III. 4. 1) is said to be the *t-sub-Planck component* of m. The inverse limit of the sequence of (III. 4. 2) is said to be the *right t-sub-Planck covering* of V, and the direct limit of (III. 4. 3) is said to be the *left t-sub-Planck covering* of V. A presheaf in $\hat{S}$ and an object of S appearing in the t-sub-Planck component are said to be a *t-sub-Planck presheaf* of $\hat{S}$ and a *t-sub-Planck object* of S, respectively. One may regard these objects as *ultramicro-objects*.

Definition III. 4. 4 A sequence of decomposition as in (III. 4. 1) is said to be *convergent* when after a finite number of decomposition steps, all the components of a decomposition consist of only fundamental presheaves. Similarly, a sequence of a covering as in (III. 4. 2) is said to be *convergent* when after a finite number of covering steps, all the covering objects consist of fundamental objects of S.

Remarks III. 4. 5 (1) A particle is said to be a *t-elementary* particle when the associated presheaf with the particle is a fundamental presheaf. When a pair of a fundamental presheaf of $\hat{S}$ and a fundamental object of S is reified, such a pair is said to be a *fundamental pair*.
(2) Let us consider a measurement issue as the interplay between a quantum entity and a macro object in terms of decompositions and coverings notions. See also entropy, which will follow. For a microdecomposition $m = \prod m_j$ of m, consider the projection $m = \prod m_j \xrightarrow{p_j} m_j$ in (III. 4. 1). Suppose that m_j is measured by P over V. Namely, we have a morphism $m_j(V) \xrightarrow{(S_j)v} P(V)$. Then one may not compose this observation morphism S_j with the projection morphism p_j over V, since m may not be reified over V. Namely, the morphism

76

$m(V) = (\prod m_j)(V) \xrightarrow{(p_j)_V} m_j(V)$ may not be defined which is to be composed with the morphism $m_j(V) \xrightarrow{(S_j)_V} P(V)$. This is the general difficulty of going from local data to global information. In other words, in general, for a given morphism $Q \longrightarrow P$ of presheaves Q and P, a functorial (canonical) morphism does not exist unless both Q and P are defined over the same object of the t-site. Even if one has local data from the measurements on $\{m_j\}_{j \in J}$ for all j via $m_j \xrightarrow{(S_j)} P$ over appropriate generalized time periods, one cannot measure the global object m unless the observation morphism S_j is composable with the projection morphism p_j. If m is a fundamental presheaf, then the consequence is trivial since m has only a trivial decomposition. On the other hand, for a covering $\{V \longleftarrow V_i\}$, suppose that m is measured by q over V. Namely, we have a morphism $m(V) \xrightarrow{r_V} q(V)$. There is no reason to assume that $m(V_i)$ is defined, unless V itself is a fundamental object. Even for a reified pair $m_j(V_i)$, we need not have the reified $q(V_i)$. Consequently, we cannot have a morphism $m_j(V_i) \longrightarrow q(V_i)$. This is the difficulty going from a global entity to a micro entity. See Remarks III. 3. 6 for another description on the issue of the global and the local. On the other hand, from a local entity to a global entity, we can summarize the difficulty in the following diagram. Namely, first, for reified $m(V)$ and $m_j(V_i)$, $m(V_i)$ and $m_j(V)$ need not be defined to obtain the commutativity of the diagram below.

$$\begin{array}{ccc} m(V) & \longrightarrow & m(V_i) \\ \downarrow & & \downarrow \\ m_j(V) & \longrightarrow & m_j(V_i) \end{array} \qquad \text{(III. 4.4)}$$

Secondly, the lack of commutativity of the diagram also indicates (or measures) the difficulty of observation the local $\{m_j(V_i)\}_{i,j}$ to the global $m(V)$. The difficulty from the global to the local is in principle impossible in the categorical sense because of the impossibility of composing the morphisms for the measurement morphism $m(V) \longrightarrow q(V)$ in the above diagram. The above observation between the global and the local is possibly relevant to Riemann's comments *"But if this independence of bodies from position does not exist, we cannot draw conclusions from metric relations of the great, to those of the infinitely small."* in Epitome III. 1.

(3) Next consider beyond such an ultramicro-object case defined by the inverse and direct limits. In other words, such a domain is a microcosm which cannot be reached by the limits of direct and inverse systems. This is the state where every presheaf essentially is t-virtual (or more precisely such a sequence of linearly t-ordered morphisms terminates). Even if a presheaf is not t-virtual, such a state of the presheaf has no connections to outside ultramicrocosm since there does

not exist a sequence connecting to macrocosm. One can consider the state where nothing is reified, i.e., there are no matches of presheaves and objects of the t-site. All we can tell in this formulation is that it is the realm where there are no sequences connecting from a macrocosm. See III. 5.

Epitome III. 4 The formulation gives some indications in terms of t-topos what the fabric of space-time is made of in very small scales. The five axioms (t-0) through (t-4) and their associated concepts for our t-topos theory provide basic ideas in terms of presheaves for matter and space-time in microcosm and macrocosm. When a macro object, represented by a presheaf m, is observed during the time period represented by a t-site object V, a naïve question "What's happening to micro (sub) objects $\{m_j\}_{j \in J}$ of which macro object presheaf m consists?" can be asked. Half of this question has been formulated already in Remarks III. 4.5, and the remaining part of this question will be provided in the following section. Axiom (t-4) on a covering and the associated time sheaf gives a more detailed description on such subpresheaves $\{m_j\}_{j \in J}$. When some of $\{m_j\}_{j \in J}$ are reified to be in ur-particle states with t-site objects $\{V_i\}_{i \in I}$ of a covering of V, the globally measured time period $\tau_m(V)$ of m is not a shorter period than $\tau_{m_j}(V_i)$ for each m_j. Namely, axiom (t-4) implies that the finer the presheaf m is decomposed, the less stable for subpresheaves of m to exist in particle ur-states, i.e., more likely in an ur-wave state. See following Section III. 5 for the connections of the above comments to the t-entropy concepts. Next, let us focus on a vacuum. The vacuum can be considered as being filled with small quantum fluctuations in the t-topos sense as well, i.e., undulations of fields, namely filled with only briefly living t-elementary particles presented by $\{m(V)\}_{m \in Ob(\hat{S}), V \in Ob(S)}$, where m and V are regarded as ultra micro objects in the sense of Definition III. 4. 3. Thus the vacuum is considered as consisting of ultra micro objects coming into existence for short periods before disappearing. The energy of a vacuum cannot be asserted to be zero locally (and therefore globally) producing the above fluctuating rest mass and energy coming from Heisenberg's uncertainty.

Section III. 5 Limits, t-Entropy, u-Singularities, Light Cones, and Black Holes

Capturing the nature of light has been a motivational goal in quantum physics and the relativity theory. We will define the notion of a light cone in terms of an associated presheaf γ to a photon. We have introduced for a linearly t-ordered morphism $V_1 \xrightarrow{g_1} V_0$ the notion of ur-states of a photon associated presheaf γ over V_1 and V_0. That is, as in $(t-2)$ of Definition III.1.1, γ is reified over V_1 first and later over V_0. We have the linearly t-ordered morphism $V_1 \xrightarrow{g_1} V_0$ for γ. Continue the process of linearly t-ordered morphisms obtaining the following sequence $V^\bullet$ in the t-site S:

$$V^\bullet = ---\to V_2 \xrightarrow{g_2} V_1 \xrightarrow{g_1} V_0 \xrightarrow{g^0} V^1 \xrightarrow{g^1} ---. \quad \text{(III. 5. 1)}$$

Definition III. 5. 1 Let γ be the presheaf associated with a photon. Then the sequence

$$---\leftarrow \gamma(V_2) \xleftarrow{\gamma(g_2)} \gamma(V_1) \xleftarrow{\gamma(g_1)} \gamma(V_0) \xleftarrow{\gamma(g^0)} \gamma(V^1) \leftarrow --- \quad \text{(III. 5. 2)}$$

is the *t-world line* of the photon presheaf γ for the sequence $V^\bullet$ of linearly t-ordered morphisms.

Definition III. 5. 2 Let γ be an photon presheaf reified over V_0 of the t-site S. Then consider all the possible t-world lines of γ reified with all the sequences of linearly t-ordered morphisms as in (III. 5. 1). The collection of all the sequences of the type (III. 5. 1) with vertex at V_0 is denoted as $\{V^\bullet(V_0)\}_{V^\bullet \in Seq}$, where Seq is the totality of all the sequences as in (III. 5. 1) passing through V_0. Then the *universal light cone* with vertex at V_0 is defined by the totality $\{\gamma(V^\bullet(V_0))\}_{V^\bullet \in Seq}$.

Remarks III. 5. 3 (1) Because of the discreteness of the existence of t-site objects in a sequence like (III. 5. 1), there exist non-reified states in the t-world line of any particle. Hence, there are also non-reified states on the universal light cone. In other words, if a sequence as in (III. 5.1) consists of linearly t-ordered micromorphisms, then between the ur-particle states $\gamma(V^i)$ and $\gamma(V^{i+1})$, $i \in \mathbb{Z}^+ = \{1,2,---\}$, where $V^i = V_i, i \in \mathbb{Z}^- = \{0,-1,-2,---\}$, the photon presheaf γ is in an ur-wave state.

(2) For any presheaf m associated with a particle, in the classical diagram for a light cone, the t-world line $m(V^\bullet(V_0))$ is contained in the universal light cone $\{\gamma(V^\bullet(V_0))\}_{V^\bullet \in Seq}$ with vertex V_0.

(3) As we mentioned earlier in $(t-2)$ of Definition III. 1.1, a particular occasion assigns a particular object V in the t-site. The true nature of the history of a particle within the light cone is needed to be phrased at a generalized time period V_0 at which that presheaf is possible to be reified. Namely, our definition of a universal light cone is just a concept. In this sense, we have chosen the terminology *"universal"* as such an ideal (potentially unrealistic) conceptual notion of a light cone.

(4) Let e be the presheaf associated with an electron. Suppose that at a non-virtual object V_0 of the t-site S, e was observed by P. Namely, we have an observation morphism $e(V_0) \xrightarrow{S_{V_0}} P(V_0)$. After the observation, the presheaf e is in a *superposition* in the following sense. The presheaf e is in an ur-wave state with the

potential of being in an ur-particle state within the universal light cone with the vertex at V_0 as defined in the above.

(5) In general, one needs to answer the following question. How many linearly t-ordered morphisms (or micromorphisms) are there from such a non-virtual object V as in Remark III. 4. 5(4)? Our approach (ontology) says that a particular object of the t-site determining an ur-state of an entity is assigned for a particular moment (occasion). That is, the number depends upon such a t-site object V. Let $V \to V^1$ be a linearly t-ordered morphism where the observation was made at V^1. In other words, V^1 is one of the candidates for such linearly t-ordered morphisms from V. Notice the following crucial assertion. If $V \to V^1$ and $V \to {'V^1}$ are both linearly t-ordered *micromorphisms*, then by the very definition of a micromorphism, there does not exist a linearly t-ordered morphism from either ${'V^1} \to V^1$ or $V^1 \to {'V^1}$. Note, however, that there can be a non-linearly t-ordered morphism either from $V^1 \longrightarrow {'V^1}$ or ${'V^1} \longrightarrow V^1$, which is neither in the universal light cone for V^1 nor ${'V^1}$. In terms of a presheaf m reified at V, V^1, and ${'V^1}$, for example, the induced morphism $m(V^1) \leftarrow m({'V^1})$ does not exist. This assertion means that the two ur-states of m corresponding to the generalized time periods V^1 and ${'V^1}$ are independent. In other words, no information can be exchanged between the ur-states $m(V^1)$ and $m({'V^1})$. However, when one of those morphisms from V, e.g., $V \to {'V^1}$ is not a micromorphism, then there may exist a morphism $V^1 \to {'V^1}$ so that the diagram

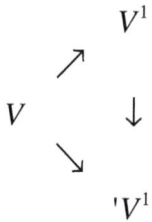

may become commutative. One may study how to formulate the connections between the possible linearly t-ordered morphisms from V and the notion of a *multiverse*. One asks for an appropriate postulate for such a number of linearly t-ordered morphisms from V for such a situation. Does the number of linearly t-ordered morphisms from V depend on not only V but also upon a presheaf m? More explicitly, for a linearly t-ordered morphism $V \xrightarrow{g} V^1$, one can ask how many factoring sequences of g by linearly t-ordered morphisms there are between V and V^1.

We do not exclude the study of the physical meaning of the case where the observed and the observer are not in a universal light cone. However, as is often the case especially with a macrocosm, we need to consider the following case when an observed presheaf m and an observer P are mutually in a universal light cone. We will give a definition of the t-topos theoretic relativistic observation (measurement) of an object m by P in $\hat{S}$. Namely, this non-local macrocosmic situation is exactly where the same generalized time period V may not be used, e.g., in (3) of Remark III. 1. 2. Let $m(V)$ be an ur-particle state of m over V and let $P(V')$ be an ur-particle state of P over V'.

Definition III. 5. 4 Two states $m(V)$ and $P(U)$ of m and P in $Ob(\hat{S})$ are said to have *mutually intersecting universal light cones* if there exists a generalized time period Z together with linearly t-ordered sequences $V^\bullet$ and $U^\bullet$ so that
$$V = V^0 \to V^1 \to ---\to V^N = Z = U^{N'} \leftarrow ---\leftarrow U^0 = U.$$

Notes III. 5. 5 (1) Then the t-world line of m is contained in the universal light cone with the vertex V^0. The case where $Z = U$ is treated in [40].

(2) For $m(V)$ and $P(V')$, if there exists a sequence $V^\bullet$ of linearly t-ordered morphisms connecting V to V', by composing those morphisms of the sequence $V^\bullet$, we get the morphism $m(V) \longrightarrow P(V')$. This means that P can receive information at V' about the past state of m corresponding to V. This non-canonical morphism $m(V) \longrightarrow P(V')$ is a generalization of Remarks III. 1. 2. (4) to the relativistic case. Then $m(V)$ and $P(V')$ are said to be mutually in a light cone. See [40]

Entropy

The t-topos theoretic notion of entropy was introduced in [43]. Our fundamental approach to this concept begins with a reified state $m(V)$, a decomposition $\prod_{j \in J} m_j$ of m, and a covering $\{V \leftarrow V_i\}_{i \in I}$ of V. We focus on the number of the reified pairs $\{m_j(V_i)\}_{j \in J, i \in I}$ among all the possible $Card(I \times J)$ pairs, where $Card(I \times J)$ is the cardinality of the set $I \times J$. We will give definitions of entropies for a given state of a reified pair $m(V)$ of m in $\hat{S}$ over V in S. We may assume that $m = \prod_{j \in J} m_j$ is a microdecomposition of m, since, if necessary, we can further decompose m_j as $m_j = \prod_{j \in J} m_{j_k}$. Let $\{V \longleftarrow V_i\}_{i \in I}$ be a microcovering of V.

Definition III. 5. 6 The *t-entropy* of an ur-particle state $m(V)$ with respect to a microdecomposition $m = \prod_{j \in J} m_j$ and a microcovering $\{V \longleftarrow V_i\}_{i \in I}$ of V is defined by the number of compatible (reified) pairs $\{m_j(V_i)\}_{j \in J, i \in I}$ among the presheaf components of m and the covering components of V, respectively. We denote the t-entropy of $m(V)$ as $(t - Ety)_{m(V)}$.

Definition III. 5. 7 The *formal entropy* of $m(V)$ with respect to a microdecomposition $m = \prod_{j \in J} m_j$ and a microcovering $\{V \longleftarrow V_i\}_{i \in I}$ is defined by $Card(I \times J)$, the cardinality of the product of the index sets J and I. We denote the formal entropy of $m(V)$ by $(f - Ety)_{m(V)}$.

Remarks III. 5. 8 (1) Let $V \longrightarrow U$ be a linearly t-ordered micromorphism, and let $\{U \longleftarrow U_i\}$ be a covering of U. Then the fibre product

$$\begin{array}{ccc} U_i \times V & \longrightarrow & U_i \\ \downarrow & & \downarrow \\ V & \longrightarrow & U \end{array}$$

gives a covering $U_i \times V \to V$ of V. Even when $m_j(U_i)$ is reified, $m_j(U_i \times V)$ may not be reified. See Definition I. 3. 1. This formulation is the t-topos aspect of non-decreasing entropy for a linearly t-ordered morphism $V \longrightarrow U$.

(2) In [43] we also introduce the notion of *the absolute entropy* for the ur-state $m(V)$ as *the maximum number* of reified pairs for all the decompositions of m and the covering of V. We are not able to know which one is most appropriate for describing the measurement of the randomness of the system given by decompositions and coverings. However, by any one of the definitions given in terms of the numbers of the reified pairs of $\{m_j\}_{j \in J}$ and $\{V_i\}_{i \in I}$, a prestage preceding the big bang, the entropy of this trivial case for the definitions becomes zero. See the section on Hypothesis on Ur-Bigbang of the xth Stage in what will follow.

(3) As a special case, the t-entropy of a fundamental presheaf is either 1 or 0. For a system consisting of two fundamental presheaves, the t-entropy is 0, 1, or 2. Let m be a fundamental presheaf and let $V \longrightarrow V'$ be a linearly t-ordered morphism. Further assume that V is a fundamental object of the t-site. For a covering $\{V' \longleftarrow V'_i\}$ of V', consider the following diagram.

$$\begin{array}{ccc} V'_i \times V & \longrightarrow & V'_i \\ \downarrow & & \downarrow \\ V & \longrightarrow & V'. \end{array}$$

Even when m and V'_i are reified, m and $V'_i \times V = V$ need not be reified. This is the local micro version of the above Remark III. 5. 8 (1). Notice that we have $V'_i \times V = V$ since V is a fundamental object of the t-site. Note also that $V \longrightarrow V'$ need not be a micromorphism, but t-ordered linearity is needed.

For the space-time sheaf ω, recall from Definition I. 3. 2 that for a covering $\{V \xleftarrow{g_i} V_i\}_{i \in I}$, we have the exact sequence

$$\omega(V) \xrightarrow{\omega g_i} \prod_i \omega(V_i) \underset{\omega p_2}{\overset{\omega p_1}{\rightrightarrows}} \prod_{i,j} \omega(V_i \times V_j), \quad \text{(III. 5. 3)}$$

where $p_k, k = 1,2$ is the projection as indicated in Definition I. 3. 2. Namely, when the space-time state $\{\omega(V_i)\}_{i \in I}$ agrees with the products (classically overlapping intersections) $V_i \times V_j$, then each local space-time state $\omega(V_i)$ may be viewed as the image of the restriction morphism $\omega g_i : \omega(V) \to \omega(V_i)$ of the uniquely determined global space-time state $\omega(V)$. This assertion explains why space-time appears to be *globally smooth* even though the instability of microcosm locally causes fluctuation. On the other hand, as in Note III. 1. 3., for $\{m \xleftarrow{t_i} m_i\}$, the induced sequence

$$\text{Hom}_{\hat{S}}(m, \omega) \to \prod \text{Hom}_{\hat{S}}(m_i, \omega) \rightrightarrows \prod \text{Hom}_{\hat{S}}(m_i \times m_j, \omega) \quad \text{(III. 5. 4)}$$

is exact. Namely, $\{m \xleftarrow{t_i} m_i\}$ acts as a universally effective epimorphism covering in $\hat{S}$ which is regarded as the general relativity version of space-time sheaf ω affected by the distribution of particles $\{m_j\}_{j \in J}$. Or, if the site has the canonical topology, by Little Zen of Yoneda in Chapter I, Section 3, we can consider all the categories coincide $C \approx \tilde{C} \approx \hat{C}$. One can formulate such an exact sequence as (III. 5. 4) at any level one prefers whether objects are either in the site or in the topos over the site. The mathematically formal observations via (III. 5.4) may have some physical meanings and relevant consequences, or may have at least some ontological interest.

Schematic diagrams for the case where observers P and Q are mutually intersecting universal light cones are given as

$$\begin{array}{ccccccc} U_{i'} & \to & m_i & & & & \\ \downarrow \nwarrow & & \searrow & & & & \\ U & U_{i'} \times U_{j'} & & m & \xrightarrow{\sigma_m} \omega \xrightarrow{S} P & & \text{(III. 5. 5)} \\ \uparrow \swarrow & & \nearrow & & & & \\ U_{j'} & \to & m_j & & & & \end{array}$$

and

$$\begin{array}{ccccccc}
V_{i'} & \to & m_i & & & & \\
\downarrow \nwarrow & & \searrow & & & & \\
V & V_{i'} \times V_{j'} & m & \xrightarrow{\sigma_m} & \omega & \xrightarrow{S'} & Q. \\
\uparrow \swarrow & & \nearrow & & & & \\
V_{j'} & \to & m_j & & & &
\end{array} \qquad (\text{III. 5. 6})$$

The above diagrams show the inner relationship in the macrocosm case between (III. 5. 3) and (III. 5. 4), where $P \xrightarrow{E-L} Q$ is the *Einstein-Lorentz morphism* (or *E-L natural transformation*) of observer P to observer Q. See [40, p. 176]. One can also give similar schematic diagrams for the microcosm case where observers P and Q affect ω. The reader is recommended to describe explicitly such rather complex diagrams. Note that in the case of the microcosm there are morphisms from both P and Q to ω as well. Thanks to Yoneda's Lemma and its embedding, diagrams (III. 5.5) and (III. 5.6) as above are valid diagrams as if they belong to the same category.

(2) We need to consider the effect of non-reified presheaves on space-time. Namely, even though ω is the terminal object of $\hat{S}$, for an arbitrary presheaf m in $\hat{S}$ the unique natural transformation $\sigma_m : m \to \omega$ may not induce a morphism $\sigma_m{}_{V'}^{V} : m(V) \to \omega(V')$ for any V and V'. Let $m = \prod_{j \in J} m_j$ be a decomposition of m and let $\{V \longleftarrow V_i\}_{i \in I}$ be a covering of V as before. For each m_k, we have a unique morphism $\sigma_{m_k} : m_k \longrightarrow \omega$. When m_k and ω are in mutually intersecting universal light cones, there exist V and V' so that we have a morphism $m_k(V) \longrightarrow \omega(V')$. See Definition III. 5. 4 and Notes III. 5. 5. (2). That is, even if presheaves are impossible to be measured (observed) directly by an observer who is not in the mutually intersecting light cones, the space-time can be affected by the presheaves as long as such presheaves and the space-time sheaf are in mutually intersecting universal light cones. A certain number of presheaves and the space-time sheaf on which an observation is made are in the mutually intersection universal light cones. Note also that such a morphism $m_k(V) \longrightarrow \omega(V')$, if it exists, is not a canonically induced morphism, i.e., not a morphism of functors. (See Definition I. 1. 9.)

Black Holes

The dynamical aspect of our theory involves a particle representing presheaf m, space-time presheaf ω, and ur-state controlling t-site sequences of objects. Most importantly we will introduce the following hypothesis called the t-topos theoretic gravitational hypothesis (referred to as the *t-g. hypothesis*). The t-g. hypothesis implies "The stronger the gravity is, the fewer factorizations exist."

T-Topos Theoretic Gravitational Hypothesis Let m be a presheaf representing a particle possibly with mass and let $V \xrightarrow{g} U$ be a linearly t-ordered morphism of t-site where m is reified at V and U. For the ur-particle states of m, consider the case where

$$V \xrightarrow{g_0} W_0 \xrightarrow{g_1} W_1 \to ---- \xrightarrow{g_k} W_k = U$$

is a linearly t-ordered microfactorization of $V \xrightarrow{g} U$ (i.e., $\{g_i\}_{i=0,---,k}$ are linearly t-ordered micromorphisms, i.e., a *linearly t-ordered pure microfactorization* in Definition III. 3.2.) for m, without the gravity effect by a particle associate presheaf M.

Next let

$$V \xrightarrow{g'_0} W'_0 \xrightarrow{g'_1} W'_1 \to ---- \xrightarrow{g'_{k'}} W'_{k'} = U$$

be a linearly t-ordered microfactorization of $V \xrightarrow{g} U$ for m together with the gravity effect by M. Then the t-g. hypothesis asserts $k' \leq k$. This is the t-topos theoretic interpretation of the generalized time effect by gravity. The inequality $k' \leq k$ is said to be the *microzation effect* in the following sense. The particle represented by a presheaf behaves more like a microcosm entity under larger gravity.

Notice that the more massive the particle represented by m is, the smaller the t-microlength becomes. Namely, as the gravity increases, m is more likely in an ur-wave state. For an extreme case where m itself represents a black hole, the t-microlength n_m can be either 0 or $-\infty$. Note that $n_m = -\infty$ indicates the case when there does not exist any reified ur-particle state for m, and $n_m = 0$ indicates the case when a linearly t-ordered micromorphism does not have either a domain t-site object or a codomain t-site object for m.

Notes III. 5. 9 (1) Consider the above factorization $V \xrightarrow{g_0} W_0 \xrightarrow{g_1} W_1 \to ---- \xrightarrow{g_k} W_k = U$ as the case where m is the only presheaf representing a particle possibly with mass, i.e., without any effect of another presheaf M presenting a particle. The t-microlength of $V \to U$ is $k+1$ in the sense of Definition III. 3. 2. The case of a photon $m = \gamma$ is fundamental in the following sense. Let n_0 be the non-negative integer n for the factorization of $V \xrightarrow{g} U$ for $m = \gamma$ in the t-g. hypothesis. Then n_0 is said to be the *universal t-microlength* of $V \xrightarrow{g} U$ for $m = \gamma$. Namely, universal t-microlength of 1 can be used as a fundamental unit for ω so that one can call such a unit the universal Planck unit for the closest ur-particle states of ω.

(2) Let $V \xrightarrow{g} V''$ be a morphism for a particle representing presheaf m. Consider the microcosm case $V \xrightarrow{g_0} V' \xrightarrow{g_1} V''$ of t-microlength 2. By Definition III. 3. 2, $V \xrightarrow{g} V''$ is a 2-micro generalized time period morphism. Then for a large enough gravitational effect caused by M for the sequence $V \xrightarrow{g_0} V' \xrightarrow{g_1} V''$, the sequence $V \xrightarrow{g_0} V' \xrightarrow{g_1} V''$ of 2 t-microlength becomes $V \xrightarrow{g'_0} V''$. Namely, the microlength of $V \xrightarrow{g} V''$ becomes 1 by the t-g. hypothesis. Consequently, between the generalized time periods V and V'', m cannot be in an ur-particle state, i.e., always in an ur-wave state. Namely, in general, under a larger gravitational (or acceleration) effect, an entity is more likely to be in an ur-wave state.

Since our approach is to replace the classical singularities involving infinity with u-singularities expressed as direct and inverse limits defined by the universal morphism properties, there is less significance at the classical singularity at the center of a black hole. Rather we pay attention to a Schwarzschild event surface Σ and capture Σ as a pseudo-category change surface (to be defined in what will follow). We denote the inside of the Schwarzschild event surface Σ of a black hole by Σ^+ and the outside Σ by Σ^-. That is, there can exist a morphism from a presheaf m over an object of the t-site whose particle is located outside Σ^- of Σ to a presheaf n over an object of the t-site whose particle is inside Σ^+ passing through a Schwarzschild event surface Σ, but the reverse direction morphism does not exist.

Let γ be a presheaf representing a photon. Consider a sequence of linearly t-ordered morphisms and objects

$$V^{\bullet} : V^0 \to V^1 \to --- \to V^N \to V^{N+1} \to --- \quad (III. 5. 7)$$

such that $\gamma(V^0)$ is located in Σ^-, outside a Schwarzschild event surface Σ. Furthermore, let N be such that $\gamma(V^N)$ is located outside Σ^- of Σ and let $\gamma(V^{N+1})$ be inside Σ^+ of Σ. The sequence (III. 5. 7) induces the sequence $\gamma(V^{\bullet})$ as follows:

$$\gamma(V^{\bullet}) : \gamma(V^0) \leftarrow \gamma(V^1) \leftarrow --- \leftarrow \gamma(V^N) \lrcorner \gamma(V^{N+1}) \leftarrow --- \quad (III. 5. 8)$$

where $\lrcorner$ indicates the morphism which cannot be composed. We characterize a Schwarzschild event surface as the boundary for a *pseudo-category change* as in the above. That is, such a sequence as (III. 5. 7) induces the impossibility of the composition on the evaluated presheaf level as in (III. 5.8). The classical space-time singularity of a black hole may be expressed in terms of t-topos notions as the ur-stage of the space-time ω of the direct limit of the direct system (III. 5. 7).

Since the t-topos theoretic entropies are defined in Definitions III. 5. 6, III. 5. 7, and Remark III. 5. 8 (2), we will explain where the early low entropy in the initial state of the universe comes from. Our method of capturing a singularity is in terms of the concept of either an inverse or a direct limit of an inverse system or a direct system, respectively. Those notions of categorically defined limits via universal mapping properties are the unifying methods for singularities of a black hole type and a big bang type.

Hypothesis on Ur-Bigbang of the xth Stage

An application of t-topos theory to the field of quantum cosmology will be touched upon briefly in this section. Any theory with a singularity as a corollary of that theory is not complete. Possible causes for a singularity whose value is assigned an ∞ as a scale can come from the structural incompleteness and an inappropriate choice of a field of numbers associated with physical entities.

We reserve the terminology Big Bang for the classical singularity based on the standard particle and cosmology theories. In what will follow, we will make a few remarks based on previously established notions in terms of categories and sheaves, i.e., t-topos methods. The terminology an ur-bigbang is used when such a singularity-like concept is expressed based on our methods. It may not be desirable to give a mathematical formulation *before* the Big Bang and the ur-bigbang. This is because such a stage cannot be reached from the present time; in terms of our theory, it is not expressible as a limit (an inverse or a direct limit of a system). However, it seems crucial to investigate and explicitly formulate such a stage or state if one claims a theory to be relevant to a unification of classical notions of a void (corresponding to our t-topos $\hat{S}$ itself), a quantum fluctuation of the vacuum and Heisenberg's uncertainty principle. Because of the t-topos theoretic methods we have developed, what will follow is, however speculative, leading to natural and reasonable consequences. One may recognize the similarities in our study of the early (and ultra-early) universe and the microcosm (the ultra-microcosm) due to the usage of limits. Our categorical (i.e., sheaf theoretic) methods to study the (ultra-) early states of the universe and the (ultra-) microcosm have been carried out in terms of not only the notions of limits but also (linearly t-ordered) micromorphisms and microdecompositions. The following descriptions of ur-bigbang states may be considered to be natural consequences of our earlier developments of the t-topos theoretic singularities (u-singularities) and the t-entropy concepts.

The ur-states of the direct and inverse limits over sequences of linearly t-ordered morphisms and objects from the present time is said to be in the *ur-bigbang of the 1st stage*. During the early states of the universe, by t-g. hypothesis, we consider that the t-microlengths of *any* sequences for all the presheaves are either 0 or 1. We can also consider a more primitive stage of the universe that consists of only reified fundamental presheaves and fundamental t-site objects without any linearly t-ordered morphisms leading to the present time in any t-linearly ordered way. In this case, one could say that every sequence is of t-microlength zero. See Definition III. 3.2. We can call such universes in primitive stages as *ur-universes*. Using the terminology of Remark III. 3. 4, the early *ur-universe* where there exists no before-after time concept, consists of reified fundamental presheaves which are in t-virtual particle states. We call this stage the *ur-bigbang of the $(-1)^{st}$ stage*. That is, at the ur-bigbang of the $(-1)^{st}$ stage, such t-site objects, with which fundamental presheaves are reified, consist of t-virtual objects. Namely, since all t-virtually ur-particle state objects $\{m(V)\}$ are isolated, their associated light cones are degenerated. However, even in the $(-1)^{st}$ stage, we still allow the existence of non-linearly t-ordered morphisms from such $\{V\}$ with which isolated reified pairs $\{m(V)\}$ of fundamental presheaves are reified. Notice that one could also speculate: the $(-1)^{st}$ stage consists of reified presheaves with morphisms; however, those t-site morphisms are not linearly t-ordered. Namely, such a reified pair is isolated with respect to linearly t-ordered morphisms. Furthermore, as a middle stage of the above stages, there can be a state where some presheaves are representing finitely short-lived virtual particles. Namely, there may be linearly t-ordered morphisms, but such linearly t-ordered sequences terminate before the ur-bigbang of the 1st stage. One can introduce the notion of a *linearly t-ordered sequence of morphisms of finitely bounded length*. Namely, a sequence of linearly t-ordered morphisms $V^0 \to V^1 \to --- \to V^N$ terminates after a finite number of linearly t-ordered morphisms with the last object V^N being a virtual object. We call this stage the *ur-bigbang of 0th stage*. That is, all the light cones are finitely terminated. In other words, the corresponding states beyond the reach of the direct and inverse limits from the present time (i.e., localized universes in which short-lived particles without connection to this universe resided) can be considered. However, by the t-g. hypothesis, the t-microlengths of the sequences for presheaves may be either 0 or 1 during the 0th stage. Namely, the index N of the last object V^N of the above linearly t-ordered sequence is 1. Note that the transition from the ur-bigbang of $(-1)^{st}$ stage to the 0th stage is not a t-topos theoretic quantum tunneling since there does not exist a linearly t-ordered morphism connecting those stages. Needless to say, during the 0th stage consisting of sequences of finite length, a t-topos quantum tunneling can occur within the 0th stage. Similarly, a quantum tunneling effect does not exist from the 0th stage to the ur-bigbang of the 1st stage. In addition to the above three stages of ultra-ur-universes, one could consider even the most primordial stage where there do not exist any reified pairs of presheaves and t-site objects at all, which could be said to be the *ur-bigbang of $(-\infty)^{th}$ stage*. In a way, it is only the t-topos $\hat{S}$ itself without any reified pairs. The following remark on the connection between the ur-bigbang of x-th stage and the t-topos version of quantum tunneling is in order. According to our formulation, a quantum tunneling of a presheaf m can occur for a micromorphism $V \longrightarrow V'$. Our definitions of ur-universes (i.e., the $(-1)^{st}$ stage and the stage $(-\infty)^{th}$) could be understood as tunneling with exits only. Namely, there does not exist such a t-site object V as in a micromorphism $V \longrightarrow V'$. See [75], especially the fourth paragraph and compare Vilenkin's consequences from the notion quantum tunneling from nothing into a de Sitter space with no initial conditions with our notions of bigbangs with the $(-\infty)^{th}$, $(-1)^{st}$, 0th, and 1st stages.

Epitome III. 5 The view based on relativity has indicated that the universe has quite possibly begun with the classical notion of a big bang, followed by inflation together with the notion of the creation of the universe from nothing. The observations during the 1990's have supported those ideas. As was measured several years ago by WMAP of NASA, it is well known that the non-uniformity in the cosmic microwave background radiation was observed possibly caused by the fluctuation during the inflation and by *non-isomorphic* fundamental reified presheaves. Particle physically speaking, quantum gravity is believed to be able to explain such a small enough universe (i.e., less than 10^{-43} seconds from the big bang) quantum mechanically.

Our t-topos theoretic approach to the early universe is that there did not exist reified space-time without a terminating virtual t-site object. However, in our formulation we are assuming the existence of the category $\hat{S}$ of presheaves (i.e., that of t-topos), where the space-time presheaf ω is a (the) terminal object of the t-topos $\hat{S}$. That is, the reified universe was formulated from the reified presheaves with non-terminating objects of the t-site. Generally speaking, when one attempts to formulate the notion of a big bang as a beginning of a universe, our sheaf categorical formulations as presented in the above in terms of inverse and direct limits might be better suited since the classical model based on real numbers based on Dedekind-Cantor type induces a singularity. By the usage of presheaf formulation (avoiding the Dedekind-Cantor model), one may also avoid the inconsistency in the uncertainty principle of Heisenberg at the big bang.

The number of mutually non-isomorphic fundamental presheaves at the big bang seems to be more than one so that the non-homogeneity of the current universe may be induced. Note that traditionally the quantum fluctuation resulting from the non-homogeneity is interpreted as the effect by an inflation. T-topos is not capable of the implication from our current formulation of t-topos of such an event as as inflation.

Epilogue:

When a bag of water is submerged into a water pool, it is weightless. This is the trivial case of Archimedes' principle on buoyancy.

A. Schopenhauer says "Thus, the task is not so much to see what no one has yet seen, but to think what nobody yet has thought about that which everybody sees."

General aspects of the category-sheaf approach consist of morphisms in a t-topos by Yoneda's embedding the t-site. What t-topos says as a representation of nature is the following. We postulate that a temporal topos of (non-reified) presheaves exists and is then followed by reified fundamental t-virtual pairs in the sense of III. 3 and Remark III. 4. 5. (1). Some lasted finite numbers of linearly t-ordered finitely terminated sequences of objects and morphisms, but none continued to last up to the present. There are some linearly t-ordered sequences that have reached the present so that such an initial state called a big bang can be captured as the totality of inverse limits of those non-terminated linearly t-ordered sequences. J. Barbour's conceptual philosophy of a timeless universe may be relevant to our approach. Since the t-site determines the state of each particle presheaf, one could formulate the temporal topos theory without time presheaf τ. Furthermore, our gravitational effect on time is phrased in terms of the t-microlength of a sequence of the t-site, i.e., without τ. Namely, the question is whether one can formulate quantum gravity without the notion of a linearly t-ordered morphism or not, i.e., only with "the collection of moments" in J. Barbour's terminology. Furthermore, Barbour's notion of Platonia might correspond to t-topos itself (i.e., non-reified pairs of presheaves and t-site objects) so that the history of our universe is the collection of reified pairs of presheaves and t-site objects.

Finally, the following quotations from A. Grothendieck and A. Einstein seem to be relevant to our t-topos approach. In his *Reapings and Sowings* (written during 1983-1986), A. Grothendieck writes "*those probability clouds replacing the reassuring material particles of before, remind me strangely of the elusive 'open neighborhoods' that populate the toposes, like evanescent phantoms, to surround the imaginary points.*" See Notices AMS (American Mathematical Society), Vol. 51, Number. 9, Number 10, 2004, and see also Notices AMS, Vol. 57, Number 9, 2010.

On the other hand, in *Out of My Later Years* (1936) published by Philosophical Library, Inc., 1950, A. Einstein writes, "*To be sure, it has been pointed out that the introduction of a space-time continuum may be considered as contrary to nature in view of the molecular structure of everything which happens on a small scale. It is maintained that perhaps the success of the Heisenberg method points to a purely algebraical method of description of nature, that is to the elimination of continuous functions from physics. Then, however, we must also give up, by principle, the space-time continuum. It is not unimaginable that human ingenuity will some day find methods which will make it possible to proceed along such a path. At the present time, however, such a program looks like an attempt to breathe in empty space.*"

The method of our temporal topos is an attempt to see the complexity of the microcosm and macrocosm through categories and sheaves.

References

[1] Artin, M., *Grothendieck Topologies*, Mimeographed Notes, Harvard University (1962).

[2] Banica, C. and Stanasila, O., *Algebraic Methods in the Global Theory of Complex Spaces*, John Wiley and Sons (1976).

[3] Barbour, J., *End of Time*, Oxford University Press (2001).

[4] Beilinson, A. A., and Bernstein, J., *Localisation of g–Modules*, C. R. Acad. Sci. Paris Ser. I. 292, pp. 15-18 (1981).

[5] Bjork, J.-E., *Analytic D-Modules*, Kluwer Acad. Publ. (1993).

[6] Bjork, J.-E., *Rings of Differential Operators*, North-Holland Mathematical Library, 21, North-Holland Publ. Co. (1979).

[7] Borel, A. et al., *Sheaf Theoretic Intersection Cohomology* in Seminar on Intersection Cohomology, Progr. Math., 50, pp. 47-182, Birkhauser (1984).

[8] Borel, A. et al., *Algebraic D-Modules*, Academic Press (1994).

[9] Bosch, S., Guntzer, U., Remmert, R., *Non-Archimedean Analysis: A Systematic Approach to Rigid Analytic Geometry*, Grundlehren der mathematischen Wissenschaften, Vol. 261 Springer-Verlag (1984).

[10] Buttterfield, J. and Isham, C., *A Topos Perspective on the Kochen-Specker Theorem: I. Quantum States as Generalized Valuations*, Int. J. Theor. Phys., 37, 2669 (1998).

[11] Buttterfield, J. and Isham, C., *A Topos Perspective on the Kochen-Specker Theorem: II. Conceptual Aspects and Classical Analogues*, Int. J. Theor. Phys., 38, 827 (1999).

[12] Buttterfield, J. and Isham, C., *Spacetime and the Philosophical Challenge of Quantum Gravity*, arXiv:gr-qc/9903072 v1(1999).

[13] Cartan, H., and Eilenberg, S., *Homological Algebra*, Princeton University Press (1956).

[14] Dimca, A., *Sheaves in Topology*, Universitext, Springer (2004).

[15] Eilenberg, S., and MacLane, S., *General Theory of Natural Equivalences*, Trans. Amer. Math. Soc. 58, pp. 231-294 (1945).

[16] Fantechi, b., Gottsche, L., Illusie, Luc., Kleiman, S. L., Nitsure, N., Vistoli, A., *Fundamental Algebraic Geometry; Grothendieck's FGA explained*, Math. Surveys and Monographs, Vol. 123, Amer. Math. Society, (2005).

[17] Fritzsche, K. and Grauert, H., *From Holomorphic Functions to Complex Manifolds*, Graduate Texts in Mathematics 213, Springer-Verlag (2002).

[18] Genovese, M., *Research on Hidden Variable Theories: A Review of Recent Progresses*, Physics Reports, 413, pp. 319-396 (2005).

[19] Grothendieck, A., *Elements de Geometrie Algebrique* (EGA) III, Publ. Math. IHES 11 (1961), 17 (1963).

[20] Grauert, H. and Remmert, R., *Coherent Analytic Sheaves*, Grundlehren der Mathematischen Wissenschaften 265, Springer-Verlag (1984).

[21] Gelfand, S. I., and Manin, Yu. I., *Methods of Homological Algebra*, Springer-Verlag (1996).

[22] Grossman, B., *p-Adic Strings, the Weil Conjectures and Anomlies*, Physics Letters B. Vol. 197, No. 1, 2 (1987).
[23] Gunning, R. C. and Rossi, H., *Analytic Functions of Several Complex Variables*, Prentice Hall, Englewood Cliffs, New Jersey (1965).
[24] Guts, A. K. and Grinkevich, E. B., *Toposes in General Theory of Relativity*, arXiv:gr-qc/9610073, 31 Oct. (1996).
[25] Hartshone, R., *Residues and Duality*, Lect. Notes in Math. Vol. 20, Springer (1966).
[26] Hilton, P.J., and Stammbach, U., *A Course in Homological Algebra*, Graduate Texts in Mathematics 4, Springer-Verlag (1971).
[27] Hiramatsu, N., and Kato, G. C., *Urcohomologies and Cohomologies of N-Complexes,* Portugaliae Math. European Math Society, Vol. 67, Fasc. 4, pp. 511-524 (2010). (See also University of Minnesota, *Differential Hyperforms* I., Mathematics Report 82-101, by P. J. Olver.).
[28] Hormander, L., *An Introduction to Complex Analysis in Several Complex Variables*, Van Nostrand (1966).
[29] Isham, C., *Topos Methods in the Foundations of Physics*, pre-print, arXiv: qant-ph/1004.3564v1 (2010).
[30] Isham, C. and Butterfield, J., *Some Possible Roles for Topos Theory I Quantum Theory and Quantum Gravity*, Foundations of Physics, 30, 1707 (2000).
[31] Iverson, B., *Cohomology of Sheaves*, Berlin-New York-Heidelberg, Springer (1986).
[32] Kashiwara, H., *D-Modules and Microlocal Calculus*, Translation of Mathematical Monographs, Vol. 217, AMS (2003).
[33] Kashiwara, H., *Riemann-Hilbert Problem for Holonomic Systems*, Publ. Res. Inst. Math. Sci., Vol. 20, n. 1, pp. 319-365 (1984).
[34] Kashiwara, H., Kawai, T., and Kimura, T., *Foundations of Algebraic Analysis*, Princeton University Press (1986).
[35] Kashiwara, H. and Schapira, P., *Sheaves and Categories*, Springer (2006).
[36] Kato, C. G., and Lubkin, L., *Zeta Matrices of Elliptic Curves*, J. of Number Theory, 15, No. 3, pp. 318 - 330 ((1982).
[37] Kato, C. G., *A Note on Frobenius Maps on Fermat Curves for p-Adic Strings*, IAS notes 5354 (1989).
[38] Kato, G., and Struppa, D. C., *Fundamentals of Algebraic Microlocal Analysis*, Taylor and Francis Group (1999).
[39] Kato, C. G., *Elemental Principles of t-Topos,* Europhysics Letters, Vol. 68, No. 4, pp. 467-472, (2004).
[40] Kato, C. G., *Elemental t. g. Principles of Relativistic t-Topos*, Europhysics Letters, Vol. 71, No. 2, pp. 172-178 (2005).
[41] Kato, C. G. and Tanaka, T., *Double Slit Interference and Temporal Topos*, Found. Phys. Vol. 36, No. 11, pp. 1681-1700 (2006).
[42] Kato, C. G., *The Heart of Cohomology*, Springer-Verlag (2006).
[43] Kato, C. G., *U-singularities and T-Topos Theoretic Entropy*, Int. J. Theor. Phys., 49: 1952-1960 (2010).
[44] Kato, C. G., *Temporal Topos and U-Singularities*, In *The Big Bang: Theory, Assumptions and Problems,* Editors: J. R. O'Cornnell and A. L. Hale, Nova Science

Publishers Inc. pp. 169-179 (2011).

[45] Katz, N. M. and Oda, T., *On the Differentiation of de Rham Cohomology with respect to Parameters*, J. Math. Kyoto University, 8, pp. 199-213 (1968).

[46] Katz, N.M, *Nilpotent Connections and the Monodromy Theorem: Applications of a Result of Turrittin*, IHES, Sci. Publ. Math. No. 39, pp. 175-232 (1970).

[47] Koblitz, N., *p-Adic Analysis: a Short Course on Recent Work*, London Math. Soc. Lec. Note Seies, 46, Cambridge Univ. Press (1980).

[48] Lubkin, S., *Imbedding of Abelian Categories*, Trans. Amer. Math. Soc. 97, pp. 410-417 (1960).

[49] Lubkin, S., *Cohomology of Completions*, North-Holland, North-Holland Mathematical Studies, 42 (1980).

[50] Lubkin, S., *A p-Adic Proof of Weil's Conjectures*, Ann. of Math. (2) 87, pp. 105-255 (1968).

[51] Lubkin, S. and Kato, G. *Second Leray Spectral Sequences of Relative Hypercohomology*, Proc. Nat. Acad. Sci, U. S. A. 75, No. 10, 4666-4667 (1978).

[52] Mallios, A., *Quantum Gravity and "Singularities"*, Note di Matematica; physics/0405111 (2006).

[53] Mallios, A. and Raptis, I., *Finitary, Causal and Quantal Vacuum Einstein Gravity*, Int. J. Theor. Phys., 42, 1479; gr-qc/0209048(2003).

[54] Mallios, A. and Raptis, I., *Finitary Cech-de Rham Cohomology: much ado without C^∞-smoothness*, Int. J. Theor. Phys., 41, 1857; grqc/0110033 (2002).

[55] Milne, J.S., *Etale Cohomology*, Princeton University Press (1980).

[56] Mitchell, B., *The Theory of Categories*, Academic Press (1965).

[57] Morita,Y., *A p-Adic Analogue of the Γ-function*, J. Fac. Sci. Univ. Tokyo 22, pp. 255-266 (1975).

[58] Oda, T., *Introduction to Algebraic Analysis on Complex Manifolds*, Algebraic Varieties and Analytic Varieties in Adv. Stud. Pure Math. 1, North-Holland, pp. 29-48 (1983).

[59] Penrose, R., *The Road to Reality*, Alfred A. Knopf, New York (2005).

[60] Pitkaenen, M., *p-Adic Description of Higgs Mechanism I: p-Adic square Root and p-Adic Light Cone*, hep-th/ 9410058 (1994).

[61] Polkinghorne, J. C., *The Quantum World*, Princeton University Press (1989).

[62] Popescu, N., *Abelian Categories with Applications to Rings and Modules*, Academic Press (1972).

[63] Raptis, I., *Finitary-Algebraic 'Resolution' of the Inner Schwarzschild Singularity*, Int. J. Theor. Phys., 45, (1) (2006).

[64] Raptis, I. and Zapatrin, R. R., *Algebraic Description of Space-time Foam*, Classical and Quantum Gravity, 20, 4187; gr-qc/0102048 (2001).

[65] Riemann, B., *On the Hypotheses which lie at the Bases of Geometry*, translated by W. K. Clifford, Nature, Vol. VIII., Nos 183, 184, pp. 14-17, 36, 37, Preliminary Version: (1998).

[66] Sato, M., *Theory of Hyperfunctions*, I. II., J. Fac. Sci. Univ. of Tokyo, Sec. I. 8, pp. 139- 193, 387 - 437 (1959).

[67] Sato, M., *Sato Lectures*, delivered at Kyoto University in 1984-1985, RIMS Lecture Notes 5 (in Japanese) (1989).

[68] Stenstrom, B., *Rings of Quotients*, Springer (1972).
[69] Schubert, H., *Categories*, Springer-Verlag (1972).
[70] Smolin, L., *Three Roads to Quantum Gravity*, Basic Books (2001).
[71] Sorkin R. D., *Forks in the Road, on the Way to Quantum Gravity*, Int. J. Theor. Phys., 36, 2759; gr-qc/9706002 (1994).
[72] Street, R., *Yoneda Structures on 2-Categories*, J. of Algebra, Vol. 50. No. 2, Academic Press (1978).
[73] Tamme, G., *Introduction to Etale Cohomology*, translated by M. Kolster, Universitext, Springer-Verlag (1994).
[74] Van Oystaeyen, F., *Virtual Topology and Functor Category*, Taylor and Francis Group (2007).
[75] Vilenkin, A., *Creation of Universes from Nothing*, Physics Letters, Vol. 117B, No. 1,2 (1982).
[76] Vladimirov, V. S., Volovich, I. V., and Zelenov, E. I., *p-Adic Analysis and Mathematical Physics*, World Scientific (1994).
[77] Volovich, I. V., *p-Adic String*, Class. Quant. Grav. 4 , L83-L83; preprint CERN-TH. 4781/87, (1987).
[78] Volovich, I. V., *Number Theory as the Ultimate Physical Theory*. Preprint CERN-TH. 87, pp. 4781-4786 (1987).
[79] Volovich, I. V., *A personal communication letter*, dated March 30, 1989 (1989).
[80] Weibel, C. A., *An Introduction to Homological Algebra*, Cambridge University Press (1994).
[81] Weil, A., *Number of Solutions of Equations in Finite Fields*, Bull. Amer. Math. Soc. 55, pp. 497-508 (1949).
[82] Wells Jr. R. O., *Differential Analysis on Complex Manifolds*, Prentice Hall (1973).
[83] Weyl, H., *Philosophy of Mathematics and Natural Sciences*, Princeton University Press (1949).
[84] Witten, E., *Magic, Mystery, and Matrix, Mathematics: Frontiers and Perspectives*, International Mathematical Union (IMU), edited by Arnold, V., Atiyah, M., Lax, P., and Mazur, B., pp. 343-352 (2000).
[85] Yoneda, N., *On the Homology Theory of Modules*, J. Fac. Sci. University of Tokyo, Sec. I, v. 7, pp. 193-227 (1954).

Index

abstract differential geometry, 39
algebraic analysis, 26
algebraic geometry, 26
background independence, 55
Barbour, J., 90
Big Bang, 87
bigbang, 89
   ur-, 88
Cartan, H., 5, 25
category, 6, 25
   abelian, 25, 27
   codomain, 11
   derived, 38
   domain, 11
   dual, 7
   of abelian groups, 27
   product, 7
   sub-, 7
canonical topology, 20, 22
cohomology, 25, 29
   Cech, 30
   de Rham, 40
   1st (first), 36
   group, 25
   hyper, 40
   l-adic, 39
   p-adic, 40, 39
   sheaf, 28
   ur-, 26
complex, 31
   Cech, 31
   holonomic (and regular), 38
covering, 18, 29
   family of, 20
de Rham, 37, 39
   (right) derived, 28, 36
   hyperderived, 40
   left exact, 27
   solution, 38
direct system, 9
D-module, 26, 33, 34
Einstein, A., 90
   Einstein-Podolsky-Rosen, 63

embedding, 15
entropy
   absolute, 82
   formal, 82
   t-, 78, 82
epimorphism, 8
   universally effective, 20, 22
exclusion principle, 56
exact connected sequence, 26
Fermat curve, 40
free resolution, 34
Frobenius
   existence theorem, 38
   map, 40
functional composition principle, 59
functor, 7, 10
   constant, 48
   covariant, 9, 10, 15
   contravariant, 7, 11, 48
   representable, 13
fundamental
   object, 76
   pair, 76
   presheaf, 76, 82, 89
gamma function
   p-adic, 42
generalized time period, 18, 47, 50
germ, 10, 33, 35
globally smooth, 83
Gross-Koblitz formula, 42
Grothendieck, A., 25, 90
Grothendieck topology, 9, 18, 29, 47
holomorphic function, 33
image, 10
integrable connection, 39
inverse system, 9, 75
Isham, C., 59
isomorphism, 6, 8
light cone, 16, 44, 48
   universal light cone with vertex at, 79
limit, 9
   direct, 9, 13, 44

inverse, 9, 13, 44
   linearly t-ordered sequence, 88, 90
macrocosm, 46, 50, 54
measurement, 50
microcosm, 46, 54
microcovering, 24, 76
microdecomposition, 23, 50, 76
microfactorization
   pure, 66
   linearly t-ordered pure, 85
micromorphism, 47, 66
   linearly t-ordered, 66, 80
microzation effect, 85
monomorphism, 8, 19
morphism, 6
   Einstein-Lorentz, 84
   linearly t-ordered, 49, 50, 85
measurement, 50
   t-linearly ordered, 49
   observation, 50
   of functors, 11, 15, 44
   of sheaves, 17
multiverse, 80
mutually intersecting
   universal light cone, 53, 55, 63, 81
natural transformation, 11, 50
   equivalence, 11
   isomorphism, 11
non-Archimedean, 59
non-commutative ring, 33
object, 6
   initial, 8
   quotient, 8
   subquotient, 8, 25
   terminal, 8, 16, 48
   ultra micro, 76
   zero, 8
observation, 50
Oka, K., 25
particle ur-state, 49
Poincare lemma, 38
presheaf, 7
   fundamental, 76
   t-sub-Planck, 76
product, 23
quantum

   entanglement, 44
   fluctuation, 44, 87, 89
   gravity, 26
   tunneling, 47, 73, 88
quasi-isomorphic, 34, 35
Riemann, B., 57, 77
Sato, M., 5
scheme, 15
Schrodinger equation, 58
Schwarzschild event surface. 86
sheaf, 13, 19, 20
   associated with space-time, 16, 48
   condition, 19
   higher extension, 38
singularity, 5, 89
   u-, 87
site, 7, 18
   temporal (t-), 10, 48
spectral sequence, 36, 37
string, 39
   p-adic, 39
space-time, 13, 48
sub-Planck, 59
subquotient, 25
Teichmuller representative, 42
temporal (t-) topos, 4, 44, 48 - 50
t-,
   elementary particle, 76
   entropy, 78
   microlength, 67, 85
   sub-Planck component, 76
   sub-Planck covering, 76
   sub-Planck object, 76
   sub-Planck presheaf, 76
   site simultaneously, 64
   sub-Planck presheaf, 76
   virtual object, 72
   virtual pair, 90
   virtual particle state, 72
   world line, 79
t-g. hypothesis, 55, 84, 88
topological space, 7, 25
topology
   canonical, 20, 22
topos
twister space, 32

uncertainty principle, 44, 69, 70, 71, 74
universe
    early universe, 89
universal
    measurement presheaf, 53
    light cone, 53
    Planck length, 85
    Planck unit, 85
    t-microlength, 85
universally effective epimorphism, 20, 22
ur-
    entangled, 64
    superposition, 49

particle state, 49, 50
    universe, 87
    wave state, 49, 50
Veneziano amplitude, 41, 42
Volovich, I. V., 59
wave ur-state, 49
wave-particle duality, 51
Weil conjectures, 39
wave-particle duality, 51
Yoneda
    lemma, 14, 21, 44, 57
    little Zen of, 17, 53, 57
    embedding, 15, 57
zeta,
    endomorphism, 40

www.ingramcontent.com/pod-product-compliance
Ingram Content Group UK Ltd.
Pitfield, Milton Keynes, MK11 3LW, UK
UKHW051303180426
11947UKWH00020B/1870